HARDROCK MINING

The Hardrock Miner

Never more pride will you see in a man,
never more fear.

He's respected by his fellows,
to the boss a mere means.

His body's as hard as earth's ore;
his lungs are as frail as his job.

Loneliness follows his droning drill,
worsened by dampness and cold.

Warped stopes are as dark as his future,
echoing death with each blast.

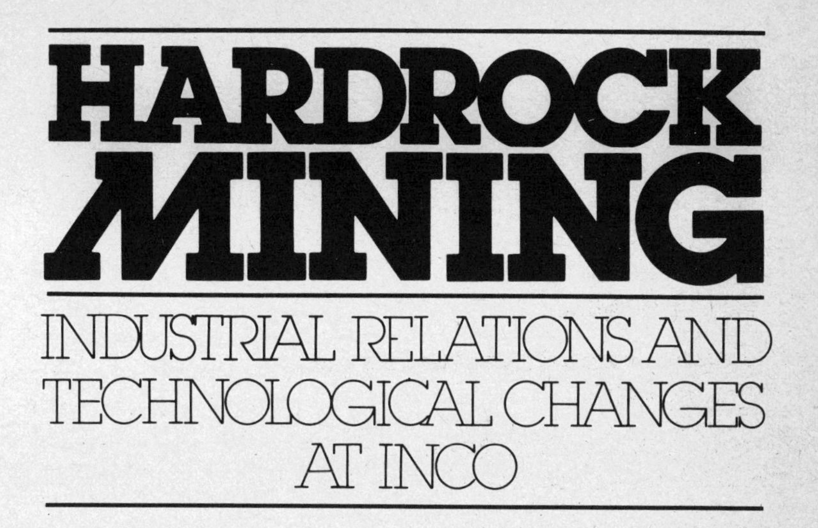

by Wallace Clement

McCLELLAND AND STEWART

McClelland and Stewart Limited
The Canadian Publishers
25 Hollinger Road
Toronto, Ontario
M4B 3G2

CANADIAN CATALOGUING IN PUBLICATION DATA

Clement, Wallace.
Hardrock mining

Includes index.

ISBN 0-7710-2152-6 bd. ISBN 0-7710-2153-4 pa.

1. International Nickel Company. 2. Nickel mines and mining - Canada - Case studies. 3. Nickel mines and mining - Canada - Technological innovations. 4. Industrial relations - Canada - Case studies. I. Title.

HD9539.N52C23 338.2'0971 C80-094399-6

Manufactured by Webcom Ltd.

Contents

PREFACE

The aims of this book are twofold. The first is a portrayal of what it is like to work in the mining industry, particularly from the perspective of those actually engaged in the work, and includes an examination of class struggles within this industry. This is accomplished through intensive examination of Inco Limited and relating it to hardrock mining in general. The second aim is longer range: it is to work *towards* an analysis, explanation, and understanding of class transformations in Canada since the Second World War. These aims are complementary. Using mining as a case study, it is possible to generate and test certain propositions about the nature of class relations, even though many of them require broader empirical testing before we can be certain that they lend themselves to generalization. But not all propositions concerning class transformations can be adequately tested through the case of mining. These are primarily aspects of class transformation concerned with the growth of employment in the service and state sectors of the economy. It is, for example, not possible to say a great deal about the position of clerical workers in the class structure through a study of mining because there are relatively few of these workers in that industry, whereas it is possible to discuss the class position of production workers, supervisors, or technical and professional workers because they are relatively numerous. Moreover, there are limitations to the "case" approach to class transformations. This method requires development of a series of studies representative of the entire society before the overall class structure can emerge, but only by making detailed examination of particular cases is it possible to develop the analysis and understanding necessary for this larger undertaking. This study is the first in a series planned to provide such an overview.

This is not a book about the laws of motion of capitalism in general or class transformations in general; that task awaits a great deal more research. Rather, it deals with their specific expression in Canadian mining, a relatively small part of the total social formation. It is hoped that through a detailed empirical study we will be closer to the larger task. The focus of this study is the structure and content of work in mining and how it has changed. It is also about the behaviour and, to a lesser extent, the attitudes of the classes involved. It attempts to go to the heart of the conditions of work and document how they are affected by managerial strategies.

It is a complex matter to portray faithfully the nature of work in Canada's mining industry. There are, in fact, two diametrically opposed views, that of management and that of workers. On the fairly narrow plane of describing the actual work techniques and technologies there is general agreement. In the interpretation of the experiences and implications of work they are usually worlds apart. There are, of course, some people within management and some workers who identify with, or at least sympathize with, the other group's perspective. For the researcher, whose own values and prejudices have no small part to play, it is easy to slip into accepting either perspective as the "accurate" one. If this happened, one would assume the world view of one side (and they are, indeed, sides) rather than examine these sides as part of one relationship. The actions of each can only be understood through the relationship of one side to the other. Yet simply presenting two views of the same occurrence will not do; we must see that there are in fact two realities and that each reality is shaped by the actions of the other.

The approach that I have adopted is to analyse various developments in the mining industry as part of a larger class relationship in Canadian society. This is the relationship between capital and labour, and it is structured in such a way that capital is granted the right to organize and direct work to expand its profits while labour, in exchange for wages, carries out the directives of capital, thus contributing to the realization of profit. For its part, management develops strategies that it thinks will maximize profits; labour, as the unequal partner in this relationship, develops means of resisting these strategies. In other words, capitalists strive to maximize the value of workers' labour power while workers seek to maximize their earnings, conditions of work, and security. At the core of the relationship is management's attempt to develop strategies that will minimize the effectiveness of workers' resistance.

While there is a constant struggle, the cards are stacked in favour of capital. Within a capitalist society, capital is the most dynamic and powerful force; it has greater resources at its disposal than labour, it is better organized, it can move its operations outside the country, and, if necessary, it can rely on the state to intervene on its side in class conflicts. Labour must sell its labour power in order to live, and the entire legitimating structure of capital and the state sanctions this activity; but labour is not without strength. When organized, it can withhold its work, albeit for limited periods of time depending on both the reserves of union treasuries and possible government back-to-work orders. It can resist the dictates of capital in the workplace by slowing down production. It can even attempt to mobilize public sentiment in its favour and bring pressure to bear on the state, although the state will not provide workers with support sufficient to keep them out of poverty unless they return to wage labour. But labour cannot, within the assumptions of capitalist society, reorganize production. This right is in the domain of management.

We are at a particularly important historical juncture for studying the mining industry. It is in the midst of a radical retooling of its basic technology, thus making possible analysis of various generations of technology and their implications for work. Underground, the direction is from hand labour to mechanization and above ground it is from batch to process technology, or what may be characterized generally as a shift from mechanization to automation. Both changes result in enormous capital expenditure and reduction in labour requirements. They have tremendous implications for the training and skills of workers and are indicative of important shifts in the forces of production. This study will explore many of the changes in the forces of production, both historically and as they are occurring today, and develop their implications for the relations of production, that is, relations between workers and machinery, among workers, and between workers and management. Like the mining industry as a whole, Inco Limited is in a state of transition. Some of its mines use traditional techniques while others have introduced heavy diesel equipment and a system of ramp mining; some of the milling, smelting, and refining plants remain much as they were at the turn of the century while others have adopted highly automated process technology to perform the same tasks. Paralleling these technological changes are important innovations in training, particularly the modular training system now being implemented with consequent reduction in the number of apprentices and skilled tradesmen. A case study of Inco can thus yield a good cross-section of past and future tendencies within this industry.

Between mid 1977 and mid 1979 Canadians heard a great deal about the layoffs and strike at Inco. There was even debate in the House of Commons about Inco's expansion abroad and the layoffs, and an Ontario parliamentary inquiry concluded that little could be done. To gain an understanding of these events it is necessary to examine in some detail various strategies Inco managers had been implementing and analyse workers' attempts to resist these strategies. The massive layoff in 1977 and strike in 1978–79 would come as no surprise to those who have examined the company closely. They are part of a longer and larger pattern. This book attempts to distil and understand the broad political-economic forces at work in the industry and reveal their implications for everyone working in the mining industry.

A detailed discussion of the methods of gathering data appears in Appendix I, but in brief, we conducted and recorded interviews with over fifty people working in the mining industry and discussed specific aspects of mining with literally hundreds of others. Both my research assistant John Baker and I were free to talk to anyone as we went through the mines and surface operations of Inco at Sudbury, Thompson, Port Colborne, and Sheridan Park and established "after hours" contacts with people we wished to interview further. Both Inco and the United Steelworkers of America were most cooperative in providing access to sites and individuals and information, although the company became more hesitant to provide assistance as the layoffs and strike developed in late 1977 and 1978. By this time, however, we had completed our tours of all the facilities. It was mainly the people in the mining industry, who gave freely of their time, whether staff or hourly paid workers, that made this study possible. These individuals spoke frankly and were willing to explain their experiences to us. They welcomed us warmly and told us a great deal. I only hope this book gives back to them some of the knowledge they shared with us.

During the course of the research I was often asked by workers and managers to explain what I was doing. Many demanded a justification. They asked what would come of the information I was gathering, what difference it would make. Who would benefit from it? These were very difficult questions to answer. Often I said that the material would be gathered together in a book or books where I would attempt to explain and make sense of what is happening in the world of work. This did not really answer their queries. When pressed, I tried to justify my work by saying it would let workers know about the feelings and experiences of other workers with whom they seldom had contact. I also said that I am trying to explain the work world in terms of broader social changes, such as the consequences of corporate concentration, capitalization, rationalization of work, fluctuations in markets, and technological change. In this way I hope to advance my own understanding and that of others about how these things affect the nature and quality of work. Basically, I am trying to understand what is so that I can contribute to what can be. I am attempting to uncover something of the human costs of work and point to the human potential.

As I have said, many people made this book possible. It is not appropriate to name them all because a promise of anonymity was an important spur to their frankness. The presidents of each of the Steelworkers' locals at Inco, Dick Martin, Ray Moreau, and Dave Patterson, were particularly co-operative and extremely welcoming. Dave Patterson was especially helpful to us after the initial stage of research when the layoffs and strike occurred at Local 6500. Tom Canning of Inco arranged visits to the plants and mines; his assistance and that of the mine and plant managers, personnel people, and engineers made possible a thorough and comprehensive survey of the company's operations. The staff at the Archives of Ontario, the Mills Memorial Library at McMaster University, and the library of the National Office of the United

Steelworkers put invaluable sources of information at my disposal. I am also grateful to the Canada Council for funding this study and to McMaster University for providing some release time from teaching to complete its writing.

John Baker was an integral part of this research. He did much of the library work and interviewing and was a welcome companion for an exchange of ideas as our joint understanding of the mining industry grew. At McClelland and Stewart, I wish to thank Peter Saunders and Peter Milroy for their continuing interest and assistance. Janet Craig has once again acted as my editor and has helped me greatly in the presentation of this material. As before, Chris and Jeff Clement have been generous in their patience with my many weeks away. Often under the pressure of time and putting aside more important pursuits, Elsie Clement has typed and re-typed this manuscript and given me encouragement to see it through.

CHAPTER ONE

Class Transformations in Mining

Class relations in mining are found within two contexts. One of these is the group of broad transformations that have taken place in property relations with the shift from the petty commodity mode to the capitalist mode of production. The other is the more specific transformation that has occurred within capitalism. Each change follows the logic of capitalist development and contributes to a dynamic relationship between the forces and social relations of production. As capitalist relations emerged they dominated and destroyed petty commodity relations and reorganized production by increased capital expenditures and a more highly developed division of labour. Not only does capitalism destroy any prior mode of production but in the course of its own growth it also transforms production from skill-based to machine-based techniques for which it creates highly developed instruments of production. Capitalists are driven, in their search for profits, to develop strategies to assure their rates of return on investment and minimize the resistance of workers. This chapter will provide a context for explaining these developments.

The Staples Approach

The most thorough assessment of the development of Canada's resources, and particularly mining, can be found in the work of Harold Innis.* His staples thesis is a statement about distortions caused in marginal societies by concentration on extracting natural resources for metropolitan markets. As he argued in the conclusion to *The Fur Trade*: "Agriculture, industry, transportation, trade, finance, and governmental activities tend to become subordinate to the production of the staple for a more highly specialized manufacturing community."[1] For Innis it was the character of the staple that determined the techniques needed to exploit it, such as the kind of capital structure, demand for types of labour, transportation systems, and production methods. Contrary to Innis, it will be argued here that capital formation and the resulting class relations explain the situation better than does the physical quality of the mineral being extracted.

It is not necessary here to retrace the succession of staples in Canadian economic development, but it is important to establish that central to the staples thesis is the historical relationship between staples. Innis wrote: "Each staple has its own peculiar developments and its peculiar relations with other staples."[2] He argued, for example, that mineral production was "in some sense a by-product of wheat production. Railways built to produce and transport wheat were responsible for the discovery and development of minerals."[3] Mineral production helped offset the costs of railways and stimulated agricultural production. These "inter-relations" provide the focal point of his mining study. As Innis said, his purpose was "to outline the effects of mining on railways and on the Canadian economy generally, and to suggest determining factors including overhead costs, hydro-electric power, technology, and the character of ore bodies."[4] This is one of the great strengths of Innis's work and the staples thesis generally: its insistence that each "case" be integrated with a multitude of other factors, not viewed in isolation. Indeed, the insistence that Canadian development cannot be understood without examining its capital, technological, and market relationships with metropolitan nations has become the common link between the

* For a more thorough exposition of Harold Innis's work on mining and a critique of his approach, see Wallace Clement, "Class Transformations in Mining: A Critique of H. A. Innis," in *The Legacy of Harold Innis*, ed. Mel Watkins (Toronto: University of Toronto Press, 1980).

original and contemporary analysts of staples. There is disagreement, however, about how these factors relate to one another. For example, Innis argued that "with a high degree of mobility of labour, finance, and technique the character of the ore body assumed a dominant position."[5] But the problem must surely be to explain how and why labour, capital, and technology become mobilized, that is, to understand the social relations between these factors.

For Innis, labour and capital are to be analysed as part of the forces of production, not in terms of their relations with each other. They both tend to enter his analysis as extensions of technology. In *Settlement and the Mining Frontier*, Innis viewed labour from the perspective of capital, that is, simply as a cost of production that is highly variable, subject only to movements of supply and demand. The analysis to be developed here, on the contrary, views staples as raw materials that capitalists wish to turn into capital. (They do not, of course, do this themselves. It is done by labour.) In the first stage, staples are treated as commodities, gathered or produced by someone else, and exchanged by merchant capitalists. With the introduction of capital directly into the production process, as is characteristic of industrial capitalism, the work is performed by wage labour. This pattern then becomes the dominant relationship in the economic system.

Differences between the dominant modes of production (petty commodity and capitalist) explain changes in the types of materials that become staples. Mining, along with the forest products industry, marks a critical break in Canada's staple production. Fish, fur, and wheat were *commercial commodities*, geared to consumption, and were gathered or produced by independent commodity producers. The producers were dominated in the market, through trading relations, by merchant capitalists who in turn sold these commodities in Europe. Mining and forest products mark the advent of *industrial raw materials* and penetration by industrial capitalists. The early stages of both these industries were characterized by independent commodity production – the gathering of gold in placer mining and the square-timber trade in forest products – but each one changed rapidly as the products of the mines and forests became integrated into industrial production, particularly for markets in the United States. The result was a destruction of independent commodity producers and the creation of wage labourers, although some traditional practices from earlier modes of production, such as the contract system, persisted.

In an extraordinarily perceptive article, Innis and Ratz outlined the profile of changing labour requirements between the different modes of production: "Labour has gradually shifted in the main from an individualistic basis to a share system prevalent in the early stages of development of basic industries, shown in the fishing industry, the fur trade, the lumber industry, placer gold mining (the lay system), and in agriculture (share farming), and to a wage system, the trend varying with the importance of capital equipment."[6] But Innis showed these transformations of labour simply as occurring alongside technological changes and not in terms of a dynamic social relationship between capital and labour.

There is a dynamic in Innis's work, but it is primarily an external dynamic, that is, the relationship between Canadian staples and external markets. The missing dynamic, and one that complements the external one, is internal class relations. One should not abstract Canadian class relations from international relations for either capital or labour. The capitalist class in Canada is clearly related to international capitalism through both capital and markets.[7] Similarly, the working class in Canada has been built on immigration and organized, especially in the resource industries, by United States-based unions.* Thus one needs to integrate the external and internal dynamics – the struggle between classes in Canada and the

* So-called international unions have had an important role in the Canadian labour movement. See Roger Howard and Jack Scott, "International Unions and the Ideology of Class Collaboration," R. B. Morris, "The Reverter Clause and Break-Aways in Canada," and Charles Lipton, "Canadian Unions," in *Capitalism and the National Question in Canada*, ed. Gary Teeple (Toronto: University of Toronto Press, 1972); Mel Watkins, "The Trade Union Movement in Canada," in *Canada (Ltd.)*, ed. Robert Laxer (Toronto: McClelland and Steward, 1973); Robert Laxer, *Canada's Unions* (Toronto: James Lorimer, 1976); Robert Cox and Stuart Jamieson, "Canadian Labour in the Continental Perspective," in *Canada and the United States*, ed. A. Fox, A. Hero, Jr., and J. Nye, Jr. (New York: Columbia University Press, 1976).

struggle of Canada within the world system. The forces of production – capital, markets, and technology – are used by Innis to explain the relationship between Canadian resources and the demands for these resources, but it is equally important to analyse how these factors condition relations of production in the movement from petty commodity production to capitalist production and within capitalist production from entrepreneurial to corporate capitalism.

The staples thesis places paramount importance on the technology necessary to extract raw materials from nature and move them to market. It is equally important, however, to recognize that the ensuing "technical division of labour" is infused with relations resulting from the "social division of labour."[8] Conflict "between the material development of production and its social form" is the result.[9]

The Class Approach

The forces of production correspond to the relations of production in such a way that capital dominates labour and uses technology and the organization of work to reinforce this control to facilitate capital accumulation. Production is both a technical and a social process where the social dominates the technical. Decisions governing the introduction of technology are determined by profitability (a social imperative for capitalists) and by the workers' willingness to accept them. This relationship is well illustrated in the mining industry. There has been a rapid introduction of sophisticated technology and an increase in labour productivity. At the same time there is evidence of dissatisfaction on the part of workers. Advanced technology, as will be evident, is an instrument the employer uses to expand productivity by labour and to regulate the labourer. This is evident not only in the equipment side of technology but also in the training side. Thus in many instances the workers have less control over the work process and ultimately their work requires lower skill levels. The discipline of labour is facilitated by a "detailed division of labour," which when introduced into the factory and administrative structures results in a breaking down of tasks through planning and control and the assignment of workers to repetitive parts of the work process, thus turning workers into "detail labourers." Most labour becomes separated from special knowledge or training and becomes simple labour with little skill content. There is also evident a reorganization of work that separates its "mental" and "manual" components so that work comes to be executed in the plant or mine and planned or conceptualized in the administrative structure. There is clear evidence that labour is transformed from a skill-based to a science-based activity. This is related to the actual process of work: the decline of skill and the rise of detail labour. At the same time management appropriates greater control over the surveillance of the work process. Traditionally, mining has been an industry where workers have had a great deal of control, both underground and on surface, but it has now been dramatically reduced for many of them.

Transformations in modes of production occur in stages. The first is "formal subordination," to use Guglielmo Carchedi's term, of labour to capital, whereby the labour process is transformed into wage labour without change in the technical conditions of production. Former artisans are brought together under one roof where they continue their prior labour processes but become subject to the property relations of capitalism. In the first stages of the transformation from petty commodity to early capitalist relations skilled workers were brought into the employ of capitalists but retained many of their earlier methods and practices of work, forced into this relationship by capitalists who gained control over their access to land or markets for their commodities. This is followed by what Carchedi calls "real subordination," in which a revolutionary change in the technical conditions of labour increases the division of labour and workers become "collective labourers."[10] Initially, workers use hand tools in the production process, but these are soon replaced by machinery. This is accomplished by the "decomposition of handicrafts, by specialisation of the instruments of labour, by the formation of detail labourers, and by grouping and combining the latter into a single mechanism."[11] Transformations within capitalism are mainly brought about by managerial strategies to gain greater control over the work process. This involves capitalization, a procedure in which tasks are differentiated and simplified by the introduction of mechanization. In mining this form of control was easier to establish on surface, where mechanization came early in the development of capitalism. Further evolution has led to automation on surface and mechanization underground.

As already suggested, capitalists must dominate petty commodity producers for two reasons: first, for access to their valued resources, such as land, mineral rights, or markets, and second, for access to their labour power. Thus the capitalist class continually dominates, displaces, and absorbs petty commodity producers either as labourers (wage labour) or as managers of their enterprises (salary labour) to perform some of the tasks of capital in overseeing the process of production. The petty commodity producer can be dominated in several ways – in the capital market, in the commodity market, in the capital goods market (cost of machinery), through access or rights to the use of land (the state), or by transportation charges. In Canada the state has taken a particularly active role in facilitating both domination and the rapid creation of highly concentrated corporations in command of mining.

The labour process is the actual activity involved when labour comes together with the instruments of production to transform the product. In petty commodity production and early capitalism in mining, an entire process was often performed by an individual or partners – such as the location of an ore body, the extraction of ore, and the sorting of precious metal from the rock. In the advanced capitalist mode of production the labour process becomes highly differentiated and specialized by breaking the process into its fine constituent parts so that the organization of work becomes complex and fragmented. Capitalization revolutionizes the instruments of labour through the technical refinement of the forces of production, thus greatly raising labour's productivity. The basis of the organization of production is shifted from the skills of workers to the pace of the machinery, and, as noted earlier, there is a shift from craft skills to detail labour when workers become machine operators. "Intelligence in production expands in one direction, because it vanishes in many others. What is lost by the detail labourers, is concentrated in the capital that employs them. . . . In manufacture, in order to make the collective labourer, and through him capital, rich in social productive power, each labourer must be made poor in individual productive powers."[12] Workers become concentrated in fewer work places and organized through a system of discipline into an industrial army, complete with supervision to oversee the labour process. Labourers themselves become specialized and attached to a limited aspect of the work process and particular machinery.

Technology has a double edge. There has been an assumption, dominant in industrial sociology, that increased technological sophistication means increased knowledge among and skill requirements on the part of employees. This is only partially correct. It is true that highly skilled jobs are created as productive activities become more capital intensive. There is a demand for researchers and technicians to oversee the development and operation of equipment and, to a lesser extent, for people to maintain it. But, on the other hand, as this stage also means the detailed division of labour, many tasks are sub-divided into minute operations requiring little skill or knowledge and made up of repetitive actions. The second kind of job outnumbers the first.

With capitalization comes the technology of training, which leads to the destruction of traditional skills, represented, for example, by apprenticeship programs in various trades (or what were once called "mysteries"*), and their replacement by company-controlled modular training programs. These new methods do not train individuals in trades but instead tie them to particular pieces of equipment and to particular processes, reducing their general knowledge and decreasing their marketability. Thus the processes set in motion by capitalization as expressed in the introduction of elaborate equipment and scientific training programs cut two ways. Workers become more productive because the machinery they operate is more efficient; they also become less valuable to capitalists because they are more readily trained and command lower

* Karl Marx notes, "Even down into the eighteenth century, the different trades were called 'mysteries' (mystères); into their secrets none but those duly initiated could penetrate. Modern Industry rent the veil that concealed from men their own social processes of production," *Capital* (1894; reprint ed., New York: International Publishers, 1967), vol. 1, pp. 485-86.

wages than tradesmen. It is important to note that fewer workers are required for even greater output. In nickel mining, underground the process of ore extraction has changed from primarily manual work, with little supervision and reliance on a bonus system to encourage production, to highly mechanized work and bulk mining techniques. Because of this change production miners require less skill, but there is greater need for skilled maintenance workers (although traditional apprenticeship programs are being threatened in this area also). On surface, much of the processing has been automated and the tendency is clearly away from the "craft" quality and toward process technology requiring only patrolling and maintenance work.

The literature on class has identified apparently contradictory tendencies accompanying technological change. Broadly, the two main ones are "embourgeoisment" and "proletarianization." Embourgeoisment traditionally means the acquisition by members of the working class of characteristics that are typical of the bourgeoisie, particularly values and life-styles. In fact, it means acquiring high enough incomes to live in a middle-class life-style. While miners have traditionally had high incomes compared with the rest of the working class, neither their values nor their life-styles have been middle class. Indeed, they have been among the most militant of Canadian workers, and much of their income has been appropriated by the high cost of living in mining communities. There has been, nonetheless, an important expansion of professional, technical, and managerial occupations within the industry. Whether the employees in these jobs have acquired the values of embourgeoisment remains to be seen, but their existence as a category is not in doubt.

Proletarianization has definitely occurred, it is usually argued, if there is a decline in the workers' skill levels and the control they have over their jobs. This trend is of greater significance in numbers than its opposite in the mining industry and stems from the same source – capitalization. It is given concrete expression in several ways. There is a rapid increase in the proportion of workers who are machine operators underground and machine tenders on surface. Whereas miners used to work primarily by hand and controlled the entire cycle of ore extraction, in capital-intensive mining workers operate one piece of equipment and perform one phase of a larger operation. The division of labour on surface is also dramatically affected. Traditionally, the milling, smelting, and refining operations required large numbers of skilled workers; today many workers have been eliminated and replaced by automatic equipment. Maintenance functions are now specialized, and the demand for highly skilled labour in the traditional trades declines as their activities are being performed by "module-trained" workers. Harry Braverman points out that "the key element in the evolution of machinery is not its size, complexity, or speed of operation, but the manner in which its operations are controlled"[13] – that is, the basic distinction is between machinery controlled by the worker and machinery that has built into it predetermined control that makes the worker a machine-tender. Since those who own the machinery can usually determine its operation, the direction of change has been towards using machinery to control workers and the pace of work. This is Braverman's main theme: the drive for productivity and the efficiency of capital are in contradiction to the place of workers in the process of production since they mean for workers a loss of skills and of control over the labour process.

Equipment or machinery is not necessarily capitalist; it becomes so only within certain social relations. For example, a sluice box is as easily the equipment of petty commodity producers as the capital of a capitalist. It is reasonable to assume, however, that once the equipment or machinery reaches a certain scale and requires more than a handful of men for operation, such as a mechanical dredge, it can no longer be utilized under petty commodity class relations. It must either become the common property of all those using it, as in co-operative ownership (an unusual development under capitalist-dominated social relations), or (more likely) become the property of capitalists who in turn employ the labour power of others to operate it. The experience of capitalism has been that petty commodity producers are unable to withstand the competition of capitalists and are thus absorbed with most of the actors becoming proletarianized. They can then offer only their labour power for sale, rather than the commodities they produce. There may be tangential developments as well. For example, the instruments of production (equipment or machinery) are themselves products of labour by workers other than those directly engaged in mining and thus create a capitalist manufacturing system (not necessarily in the country where they will be utilized) to supply capitalist mining.

Under petty commodity production (as in all modes of production) it is necessary for the workers to obtain subsistence, such as food and shelter. In agrarian petty commodity production, the people tended to produce their own food and construct their own shelter. In mining, petty commodity producers were likely to construct their own shelter but usually relied for supplies on shopkeepers, other members of the petty bourgeoisie (who owned and ran their own stores). Under capitalist production, all these activities tend to be drawn into capitalist relations.

In Canada the historical moment of petty commodity production in mining was relatively brief. Part of the reason for the speed of its demise was that capitalist mining, which already existed elsewhere, rapidly penetrated this new activity in Canada, sometimes directly through branch plants and sometimes with the help of indigenous capitalists expanding their activities. At first simple commodity producers worked alone or in pairs, using handicraft methods to obtain precious metals. They first met capitalists in a trading relationship, exchanging precious metals for cash. The cash was then used in consumption, with those who sold supplies also turning cash into capital. Some early miners relied on savings (or deferred consumption) as a means of becoming capitalists themselves. Many failed, but a few succeeded. Most often it was outside (either central Canadian or foreign) capitalists who took advantage of the transition.

Central to the transition from petty commodity to capitalist production is the relationship between capital and technology because the transformation typically involves increased use of technology, or "stored-up" labour, to replace "living" labour. It is important to establish the nature of this relationship. Large amounts of capital are required to develop and, more important, to implement sophisticated technology in large-scale capitalist production. Since only the largest capitalists have access to such large capital pools, either generated internally or from outside financial sources, they tend to monopolize the benefits of technological advance. This undercuts the relative productivity of existing smaller firms and inhibits the emergence of others.

Social labour is created when workers are drawn together to produce as a unit, whereas individual labour takes place primarily in craft settings. Technology has the potential to socialize labour, but control over that technology by capital in order to expand itself through profitability distorts this potential by directing it toward particular ends and not necessarily toward the benefit of workers or the society.

The major classes within the dominant mode of production exist in an unequal yet complementary relationship. Both the working and the capitalist classes are based on the capitalist mode of production and continue to exist only insofar as they are related to each other. The capitalist class depends upon the surplus created by the labour of the working class; the working class relies on wages in exchange for its labour. The nature of this relationship is inherently antagonistic. It is necessary for capital to accumulate and expand if it is to survive. In order to do so it must pay workers less than the value they produce. Labour struggles with capital in order to expand its wages, to control the product of its labour, and to organize production in a socially acceptable manner. Yet capital, in conjunction with the state, is granted control over the wages, products, and conditions of labour. Labour, within capitalist assumptions, can only react to the exercise of control by capitalists, not determine control itself. It can, under certain constraining conditions, withdraw its labour power, but it cannot appropriate the rights of capital. Managers, as representatives of capital, and workers (often through their unions but also outside them) manifest concretely the class struggle inherent in organizing capitalist production.

CHAPTER TWO

The Unfolding of Canadian Mining

The traditional miner was a petty commodity producer; often he worked as a prospector, a lumberjack, a farmer, or a fisherman besides mining, but he worked for himself. Petty commodity production in mining was characterized by a low level of technology, little equipment, low capitalization, and easy access to markets. The capital equipment required was often not much more than sluice boxes, flumes, and ditches, and the organization of work usually involved only partnerships and small groups; the labour was basically manual labour. These were the characteristics of mining of alluvial ore bodies and placer gold, where small claims could be staked. Around the turn of the century in Canada petty commodity production in mining gave way to capitalist production. A number of factors entered into this transition. The more advanced technology and capitalization needed for underground work (rather than "grubbing") were extensive; the state allowed staking of larger claims in order to increase its royalty revenues from mining (particularly as an ore became an industrial staple); and the relationship between miners or those producing the ore and smelter owners became important, particularly as the latter extended their operations back into extraction.

In the early stages of capitalist production the mine owners extracted ore by means of the tribute or tutwork (piecework) systems in which the work was contracted by auctioning it to groups of miners who in turn were charged for their materials and paid according to the amount of ore they produced. These systems combined elements from both petty commodity and capitalist modes of production but were clearly moving toward the latter. The capitalist owned the mine and its products; the workers worked for themselves, yet they did not own the means of production (a central feature of petty commodity production). Gradually this practice gave way to wage labour, but a bonus system for the volume, weight, or amount of work done was retained. Today the bonus or incentive system remains an important remnant of this past mode of payment. Another holdover is the "loose" supervision traditional in mining; a miner is likely to see his shift boss only once or twice a shift for a few minutes. The independence of the miner, the control he has over the pace of work, and the way he organizes his work have been characteristic of the job but are now giving way before mechanized mining and bulk mining techniques.

A general review of Canada's mining industry is a good place to start the history of the nickel industry and will illustrate the transformations in modes of production. The staples and class approaches as outlined in the previous chapter will provide the analytical basis for undertaking this brief review.*

Placer Gold: British Columbia and the Klondike

The first large quantities of placer gold found in British North America were discovered in 1857 in the gravel of the Fraser River. The "golden year" of the Cariboo was 1863, when thousands of miners flocked to that region. Many had had experience in the earlier California gold rushes and were on the scene quickly. There were few barriers to entry: the land was common property, open for anyone to stake a claim, and little equipment was needed for what was basically a handicraft activity. The gold was "hand picked" by miners using pans to

* This chapter will follow, in part, Harold Innis's approach in *Settlement and the Mining Frontier* (in Canadian Frontiers of Settlement, ed. W. A. Mackintosh and W. L. G. Joerg, Toronto: Macmillan, 1936) of taking a sample of "regions with widely divergent characteristics" and deal with each during its *formative stage*. I will include a discussion of coal mining, which Innis omitted, and, unlike Innis, analyse these various mining operations in terms of their modes of production and struggles between capital and labour.

remove the mineral from the alluvial beds of waterways. It was possible to pan a ton of gravel a day, although it was arduous work. The miner simply had to shake a pan of soil and water, gradually washing the sand and gravel away and leaving the heavier particles of gold mixed with black sand. To collect the fine gold dust, the residue and a quantity of mercury were placed in a tub of water where the gold and mercury stuck together. The mercury was then squeezed out in a porous sack and the remainder roasted to leave the gold.

Later, in the Klondike, these rudimentary techniques were adapted to cope with problems of frost and underground gravel that was frozen throughout the year. After locating a pay-streak by panning, the miners, usually working in pairs along creek beds, would build a log fire to melt a hole in the frozen ground, remove the gravel, and repeat the operation until a shaft ran to the pay-streak. They would then begin to make drifts (passages running parallel) using the same log-firing method. A windlass hoist was erected over the shaft, and the ore was drawn to the surface and washed. When a drift went far enough, a wooden track was laid and a car loaded with ore was pushed by hand to the hoist.[1] A drift was usually extended on both sides of the creek bed. The working season was four to five months,[2] and normally two men could hoist a hundred buckets a day, each containing seven or eight pans of gravel. This ore was stored in a dump until spring, when water trapped by a dam above the claim could be run over ore placed in a sluice box to wash out the gold. The sluice was a trough about twelve inches wide and nine inches deep with a trap at the bottom. Access to a large volume of water was crucial to placer mining, and this need offered the first opportunity for capitalist penetration of the petty commodity mode of production with the building of flumes. As Bill Moore has shown, "petty producers were at the mercy of the flume owners for water . . . [and] capitalists became the dominant feature of sluicing and flume operations, integrating the operations."[3]

The Pacific Coast had been hit by a series of "cyclones," as Innis called them, but the most turbulent rush came at Dawson in the Yukon in 1896–98. The region surrounding Dawson in January 1898 had some 5,000 inhabitants; by mid summer there were over 21,000. The lateness of the expansion of the discovery is attributable to the difficulty of getting to the gold fields. The law required that each miner have one year's supplies (about fifteen hundred pounds) which he had to transport over the three-thousand-foot Chilkoot Pass and six hundred miles down the Yukon River. Many men lost their lives or gave up while making the attempt. Thousands of packhorses died on the trails.* Speed was essential for the individual miner because the main claims were staked in 1896–97. A speculative market in stakes soon developed, resulting in what Innis called "pool-room prospecting." After a mining exchange was established, auctions for the claims were held and, as Innis notes, this "purchase and sale of claims laid the basis for concentration and, in turn with high labour costs, for the development of the 'lay' system."[4]

The lay system was a transition between petty commodity and capitalist modes of production. If industrial capitalism is to work, a labour supply willing to work for "reasonable" wages must be available. But in the Klondike during this period it was easy for a man to move into petty commodity production and labour was scarce. The lay system was the capitalists' compromise. Under it, an alluvial claim was "let out by its owner to two or more miners on condition that the miners pay all expenses connected with the mining and washing of the gold, the owner receiving half the gross output and paying all the royalty."[5] As the capitalists saw it, the lay system was "a mere temporary expedient designed to meet the owner's want of capital and mining knowledge, an expedient too, of use in retaining labourers when labour was scarce." It was the usual means of working claims in the Klondike, accounting in 1897–98 for three-quarters of the claims[6] and common as late as 1901.

Along with mining activity went the business of supplying the miners. The demand for labour to construct "roads, railways, buildings, and saw-mills raised wages to the [then] high figure of $10 a day."[7] There was also demand for other

* Harold Innis lists four main routes to Dawson and says of them: "Limited knowledge of these routes and ignorance of details were responsible for the migration of people by all of them and for consequent hardship and loss of life" (*Settlement and the Mining Frontier*, p. 184).

tradesmen such as tinsmiths and blacksmiths. In Innis's imagery, each rush "acted as a gigantic pump, unpredictable as to time and strength of stroke, which drew enormous supplies of labour and capital to the field concerned."[8] Capitalist companies were soon established in the area, particularly in the activities of transportation, supplies, and commerce, including two bank branches installed in 1899.

With the penetration of capitalist relations, the petty commodity system was undermined. The leading edge was the transportation system in the form of a railway over the White Pass, which made it possible to deliver heavy mining equipment. This reduced by half the cost of moving a cubic yard of gravel between 1899 and 1903, and most of the reduction was in wages to labour. The lay system and the premium on the cost of labour both came to an abrupt end. The population of the Yukon fell from 27,000 in 1901 to 8,500 in 1911 and 4,000 in 1921, but not without a fight over the rights of access to the land essential to the capitalists who were monopolizing production.

During the initial rush, claims were limited to one hundred square feet, and the allocation of lands "was carried out along principles established at miners' meetings."[9] As mining companies began to acquire the claims along entire creeks, the miners brought pressure to bear on the state, but its interest was in maximizing production and royalties, best accomplished by capitalists. Hydraulic mining regulations enacted in 1898 provided for long-term leases on areas covering one to five miles along waterways, rental of $150 a year for each mile, and an annual capitalization of at least $5,000.[10] State policy thus undercut the position of the petty producers and supported capitalist production. The miners, of course, protested the new lease arrangements, arguing that they would "hand the country over bodily to a gigantic monopoly whose interest it will be to carry out their operations with Chinese or Japanese labour."[11] A royal commission was appointed in 1903 to examine the hydraulic concessions and in particular the concession of A.N.C. Threadgold. While the commissioners rejected Threadgold's concession, "he was able shortly to acquire by other methods the claims necessary for large-scale operations" and the overall trend toward large concessions was firmly established.[12] By 1906 a Placer Mining Act had been passed, but it only served to stabilize the field for capitalist firms and encourage individual prospecting, not mining. Since capitalist firms had already gained rights of control over the claims, they were free to exploit them. The transformation from petty commodity to capitalist relations in gold mining was accompanied by increased activity by the state in this industry, thus reinforcing the strength of capitalist production and in particular of large-scale capitalists capable of taking advantage of expensive technology.

The mechanics of capitalization of alluvial mining began most dramatically in the Klondike in 1900 with the introduction of dredges. A single dredge cost $300,000 in 1905, thus limiting entry to only well-capitalized firms. "Three men running the dredge handled 700 cubic yards of gravel in 20 hours, representing 'the labour of 156 men working with a shovel and pick' – a material saving with the decline of wages to $5 per day and board."[13] Further savings resulted for the capitalists from the development of hydro-electric power, which reduced the cost from thirty cents per cubic yard when gas was used as fuel to ten cents. The Yukon Gold Company had eight electric dredges by 1913, and a few large companies controlled the entire output.[14] The transition was rapid. Aided by state policies and capital equipment developed in other countries, the domination by capitalist firms was complete. Labour became wage labour, and the dominant means of extraction became capital intensive.

Coal Mining: Vancouver Island, Atlantic Canada, and the West

On Vancouver Island native people had been gathering coal from outcroppings on the beaches of the northeast coast prior to the arrival of Europeans. This resource was harnessed by the Hudson's Bay Company after some of its officers had been led to it in 1836. Initially coal was traded by native people, who transported it to the company's ships in canoes, but soon the company established permanent operations.[15] The state granted nearly all the Crown-owned coal lands on Vancouver Island to the Dunsmuir family in 1884 (to add to their already substantial holdings in coal) as a subsidy to the Esquimault and Nanaimo Railway, following a precedent established in 1868 when the Nova Scotia government granted coal lands to the Glasgow and Cape Breton Coal and Railway Company.[16] As early as 1871 miners in the Dunsmuir operation had organized to combat wage reductions and protest against unsafe work conditions. They were particularly threatened by Dunsmuir's use of Chinese labourers as strike-breakers. By 1885 Dunsmuir was employing 450 Chinese to do manual labour and about 325 others.[17] In his review of Vancouver Island coal mines, Bill Moore comments that

> like the coal mines of Nova Scotia . . . [they] were worked under a system of contract similar to the tribute and tutwork systems since the miners were paid by the ton or yard while having to pay for the powder which they used. In the late 1870s, wages were $1.20 per ton, then were slashed to $1.00 per ton, resulting in a strike by the miners; however, this was not the first, and certainly would be followed by . . . more strikes. This system of contract continued in use into the twentieth century, since it was a central issue in the coal miners' strikes in 1913 and 1917, along with the issue of union recognition. The miners had managed, however, to win by this time a minimum day rate of $3.00 per day, regardless of the tonnage produced or yardage mined.[18]

The Royal Commission on Relations of Labour and Capital (1899) concentrated in part on Maritime coal mines, and its report provides a good source of information on the social relations of production in that industry. Springhill Mines, in Springhill, Nova Scotia, employed about 1,500 people at that time, including about 150 boys. The boys, who drove horses pulling boxes of coal, loaded and unloaded boxes, and opened traps in the mine, were paid from forty-five to eighty cents for a ten-hour day. The men, however, worked on a different system. Coal cutters worked in pairs removing the coal from the face of a drift and were paid by the box. Although the company furnished the tools they worked with, the cutters had to pay for their own blasting powder (about 5 to 10 per cent of their income) and hire a loader to put the cut coal into boxes.

Besides the loaders, who were paid from the cutters' income, the miners collectively employed a checkman (paid about 1 per cent of their incomes) to oversee the company's weigher and report back to them any irregularities in the tally, since they were regularly docked for too much rock or underweight boxes. In addition, each paid a set fee for the local doctor and school. In this system time as well as contract work was involved. The company required the miners to work complete shifts, not to stop when they felt they had done enough work for that day.[19] This form of work is another intermediate form between petty commodity production and wage labour; it contains elements of each type but clearly involves formal subordination to capitalist domination. The contract system common in early mining took two forms. In tutwork the miners were paid by the amount of ground they cut and in tribute-work they were paid by the value of the dressed ore they raised. A substantially revised form of contract work remains prevalent in mining today but now takes the form of a bonus system above an hourly wage.

Saskatchewan miners in the Souris coal fields, numbering over six hundred working in a dozen deep-seam mines in 1931, were employed under conditions similar to those prevailing in Nova Scotia and Vancouver Island. Their pay, however, consisted of a daily wage plus payment for timbering and the weight of coal mined. The cost of blasting powder was deducted from their income. The nature of the coal industry in Saskatchewan, however, differed considerably from that in other regions. The work was seasonal, the mines operating only during the late fall and winter months because the major market was for domestic use, so that there was little demand outside the cold months. During the long layoffs the miners often worked on farms. In the early 1930s this pattern was upset by the depression, which lessened both demand for coal and the availability of off-season work for the miners. Moreover, in the Truax-Traer mines mechanized strip-mining techniques were introduced which required much less labour. Together these developments led the mine owners to cut wages. The ensuing struggle, led by the Mine Workers' Union of Canada, resulted in a bloody clash between the miners and the RCMP on Black Tuesday in September, 1931. The miners were crushed and their union broken so thoroughly that they remained unorganized until 1945. This conflict between capital and labour in the Saskatchewan coal fields demonstrated the bias of the state, which was to contain labour through coercion and court action while serving the interests of capital.[20]

Coal miners have been among the most militant workers in Canada and were among the first to unionize. In Nova Scotia they formed the Provincial Workmen's Association as a craft union in 1879, and it led important strikes in 1904 and 1907, giving way in 1915 to the United Mine Workers of America. The solidarity of Alberta coal miners in 1906 precipitated implementation of the Industrial Disputes Investigation Act of 1907* and the Western Federation of Miners, active since 1895, strongly supported the Industrial Workers of the World after it was formed in 1906 and the One Big Union in 1919. Conspicuous evidence of their militancy can be seen in the fact that coal miners in Canada, representing under 2 per cent of all non-agricultural workers, accounted for 42 per cent of the time lost as a result of strikes between 1900 and 1913 and 53 per cent of the time lost between 1921 and 1929.[21]

Workers in the early coal mines, not being primarily wage labour, were paid by a complex system. Moreover, most of these early mines were quite small, employing between twenty-five and forty men using hand drills and little equipment other than steam-powered drainage pumps and hoists.[22]

Lode Gold, Copper, and Zinc: The Kootenays

In contrast to the Klondike, where petty commodity production came to be dominated by capitalist relations, in the Kootenay area of British Columbia petty commodity production in the form of placer gold mining had been exhausted by the mid 1870s, well before capitalist relations were introduced. There was, however, as Innis pointed out, a link between the earlier placer mining and the development of transportation systems in the region. These transportation systems in turn facilitated

* The Industrial Disputes Investigation Act, child of Mackenzie King, imposed compulsory conciliation on key economic sectors, particularly coal mining and railways. It is the foundation of all subsequent industrial relations policies of the Canadian state. A very informative account is available in Paul Craven, " 'An Impartial Umpire': Industrial Relations and the Canadian State, 1900–11," Ph.D. thesis, University of Toronto, 1978.

the rapid introduction of capital-intensive methods of mining, providing the opportunity for lode mining (in solid rock). In 1887 a crushing mill was built by the Selkirk Mining and Smelting Company; the ore from that mill was then exported to San Francisco for smelting. Mining in Trail for copper began in the 1890s with a smelter-railway complex constructed by F. Augustus Heinze (the Montana Copper King) of the U.S. Mining Company. His British Columbia Smelting and Refining Company began operation in 1896. In 1898 it was sold to the Canadian Pacific Railway (CPR) which was after the railway rights. The CPR was interested in stimulating mining in the area "as a means of encouraging traffic in relatively non-remunerative, high cost of construction, operation and maintenance territory, and of contributing important long-haul westbound transcontinental traffic in the form of machinery and passengers from the industrial east."[23] The railway also made possible the cheap transport of coal from the Crowsnest area to fuel the smelters.

Thus at an early stage the conditions necessary for the rapid development of large-scale capitalist operations were met, facilitated, of course, by generous state subsidies to the CPR. Eastern financial interests were quick to follow up the advantage; for example, the Gooderham family of Toronto purchased a mine in 1897. The CPR began to expand rapidly, buying mine options in 1905 to complement its smelter and forming them into the Consolidated Mining and Smelting Company (Cominco) in 1906. These developments were looked upon favourably by the state. The *Annual Report* of the Minister of Mines commented in 1903: "This viewing of mining from a more strictly business standpoint is gradually tending to the elimination of enterprises which were not based upon substantial mines and which therefore could not succeed but by their very existence cast a shadow of doubt upon legitimate enterprises."[24] As Innis noted, it was necessary for Cominco "to protect itself" against the vagaries of small mining operations by assuming control of them to ensure stability.[25] Other factors also added to the toll; although the First World War drove the price for metals up, as did the lowering of the U.S. tariff on zinc concentrates in 1913, the price decline in 1919–20 meant that "only the large mines were able to operate."[26] During the early stages of the Great Depression (1929–33) when the value of Canada's exports, two-thirds of which were raw materials, fell dramatically, workers such as miners who were lowest on the resource ladder suffered most.[27] The "busts" of mining tended to eliminate all those unable to weather such storms.

Labour was never in short supply in the Kootenay region as it had been in the Klondike. The completion of the CPR and the end of petty commodity production in the Klondike created a major labour pool. Moreover, from the outset the production facilities were capital intensive, requiring little labour, and the CPR was expert at acquiring the necessary labour through immigration. Innis noted that "the overwhelming importance of capital equipment and technical management tended to weaken the position of labour from the standpoint of unionism and to accentuate mobility, particularly with the marked increase in immigration to the adjacent plains area. Settlers worked in the mines for a small stake and moved on to their homesteads. Mining like railway construction was a source of cash to large numbers of immigrants."[28]

Innis did not show that capital used immigration in other much more direct ways to control labour. As Stanley Scott observes in his analysis of this subject, immigrant Italian labourers were systematically assembled to man the smelters. He quotes Edmund S. Kirby, manager of the War Eagle Mine, as saying: "In all the lower grades of labour, especially in smelter labour, it is necessary to have a mixture of races which includes a number of illiterates who are first class workmen. They are the strength of an employer and the weakness of the Union. How to head off a strike of muckers or labourers for higher wages without the aid of Italian labour I do not know."[29] Based on previous experience in building the railway, the CPR owners now encouraged the employment of Italians in their mining operations. These immigrants accounted for 60 per cent of the mill workers and 40 per cent of the residents of Trail.[30] Initially the mode of payment in the mines was a contract system based on the number of feet drilled. This later gave way to wage labour plus a bonus system.

The mining companies fought the organization of labour by

firing the organizers and intimidating the immigrant labourers. Working under difficult conditions, the International Union of Mine, Mill and Smelter Workers was able to form Local 15 in the Cominco plant at Trail in 1916. Its first contract represented a 15 per cent increase in pay, but this "hardly dented the inflationary rate of almost forty-seven per cent since the previous raise in 1911."[31] With workmen in abundance and a relatively low demand for labour in the operations, however, the company succeeded in breaking the union in 1917 by not recalling five hundred men to work after an unsuccessful strike. Workers at this operation thereafter remained unorganized until 1938. The battles between capital and labour in the Kootenay region of British Columbia signalled at an early stage the turbulence that became characteristic of labour relations in Canadian mining.

The Kootenay region also provided an important illustration of the transformation that had taken place within capitalism itself. During the early stages of the capitalist penetration of mining there was a great deal of competition. The competitors came from two sources, the first *internal* – those who succeeded in accumulating savings from independent commodity production and transformed this money into capital – and the other *external* – through penetration by financial capital (as in the case of Inco Limited), through backward integration by industrial capital (such as smelter operators), or by extension of corporate capitalism (as with the CPR and Cominco). Each successive stage represented different forms of accumulation as capitalists restructured themselves through combination and merger to minimize the destructiveness of competition. The restructuring left the industry dominated by large companies that were better able to embark on capital-intensive operations and resist the struggles of labour.

Silver Mines: Northern Ontario

The Ontario Department of Mines noted in the *Annual Report* of 1892 that outsiders were dominating the province's mining industry and that little processing was done in the province:

> Ontario has been disappointingly slow in developing its

> mines, and what has been done has been the work not of Ontario men, but of Americans or Europeans, and has frequently been carried on in ways unsuited to our conditions. . . . No mining region can reach the highest prosperity merely by shipping its ores to other countries, and it is safe to say that until Ontario ceases to sell its ores and low grade mattes and begins to smelt and refine its own iron, steel, nickel, and copper, no great advance is likely to be made.[32]

There was no boom from placer gold mining as had occurred in British Columbia and the Yukon to set the stage for Ontario's mining. The closest parallel was the discovery of silver in Cobalt in 1903, creating a boom that finally collapsed in the early 1920s. In 1904 there were only 57 miners in Cobalt, but by 1912 there were 3,500. Initially capitalists feared that the silver finds were confined to the surface, but after 1907, when it became evident that the veins ran deeper, open-pit mines were transformed into shaft mines calling for large amounts of capital. Shaft mining also required skilled miners and a higher level of technology for sinking shafts, driving drifts, and removing ore. Most of the skilled miners were recruited from Nova Scotia coal mines, and the "backbreaking labour was performed by unskilled immigrants from continental Europe – Poles, Italians, Austrians, Hungarians and Finns."[33] A long, bitter strike over wages in 1907 caused many of the skilled miners to leave. "In the spring of 1908 unskilled workers were reported as plentiful and skilled workers as scarce. The end of the strike was an indication of the declining importance of labour and of the increasing importance of capital," Innis commented.[34] Although this may have been true of the number of labourers required, it was not of the skill levels required of the workers.

In order to control the supply of labour, the mine owners formed the Mines Free Employment Bureau in 1907

> to coordinate the hiring policies of the operating mines. . . . Its stated purpose was to advance the mining industry in the Temiskaming District, to consider changes in mining legislation, to maintain a hospital in the area, and to promote a

> spirit of co-operation amongst the various mines in the camp. In reality, however, it was a union of the leading mine managers designed to regulate hours, wages, and working conditions. By controlling both the labour supply and the type of employees that were hired, the Employment Bureau allowed the mine owners to discipline the work force effectively. [35]

The bureau proved advantageous to capital, and by 1910 virtually all mines in Cobalt were members. This was an important indicator of the corporations' need to rationalize their supply of labour for both availability and quality. It became evident to the mine owners that a supply of skilled miners to carry out lode mining was critical to their operations. It was no longer sufficient to go on recruiting the unemployed. The owners' association represented a stage in the battle between capital and labour over wages and working conditions; capitalists had been hurt by the 1907 strike and sought to contain such uprisings.

The First World War was an additional shock to the silver mining industry. At first the markets for silver were cut off, and mines had to be closed; later the price rose rapidly and the mines reopened. "The gains were not passed on to the workers whose real wages declined due to an inexorable rise in the cost of living. These conditions provided fertile ground for union organizers."[36] The Western Federation of Miners attempted to organize the workers in 1916, and its membership rose from four hundred to fifteen hundred. Through their association the mine owners countered with wage increases, but this only postponed the strike until 1919. The strike of 1919 marked the end of the silver mines of the area. The best ore had already been extracted (known as high-grading) and the price of silver again fell. The boom over, Cobalt fell victim to the bust with its inevitable outcome for mining communities – the exhaustion of ore and the end to jobs.

The impact of the Cobalt silver industry lasted longer than this suggests. As Innis pointed out, "Companies transferred their activities from Cobalt to Porcupine and began the development of such gigantic properties as Dome, McIntyre, and Hollinger."[37] Many of the miners moved with them. In these new operations much more attention was given to labour relations because the value of skilled labour had become evident through the silver experience, and the owners knew that profitability required the mines to stay open. For example, to reduce labour turnover Hollinger Mines introduced a "loyal service" system in which workers were rewarded with an hourly bonus above wages for each year as an employee. More attention was paid to living conditions, and in 1922 the company imported a hundred Cornish miners to shore up the quality of its work force.[38] Under corporate capitalism, much of the cost of production shifts from labour to capital, but the quality of the labour required for capital-intensive operations also increases, at least during the early stages. Important distinctions begin to emerge within the working class between different skill levels, and skilled workers are able to command much better conditions and pay levels from the companies. It is in the skilled trades that labour union activity has its greatest momentum, but it is a momentum where the first interest involves only a fraction of the working class. Capital-intensive industries consequently promote divisions within the working class, and management is in a position to pay much higher wages in these industries than in labour-intensive industries.

Today there are several giant Canadian-based companies specializing in mining and operating throughout the world. They include Inco, Noranda, Falconbridge, and Cominco. From the turn-of-the-century entrepreneurial stage there have emerged dominant integrated companies capable of shifting production from country to country. The position of Canadian labour is, of course, weakened as the companies are capable of supplying part of their markets from alternate facilities in the event of strikes in Canada. The Canadian labour force is also subject to the brunt of the boom-and-bust cycles of this resource-based industry and frequently experiences massive layoffs, particularly prior to contract negotiations.

Although the history of the nickel industry (to be discussed in the following chapter) has been riddled with periods of labour shortage and layoffs, it has not been as cyclical as that of the newest mining industry, uranium. Elliot Lake, the heart of Canada's uranium mines, had twelve mines open between 1954 and 1958 (Bancroft also had four) stimulated by federal government contracts of $500 million for uranium oxide. Employment in Elliot Lake rose from five hundred to ten thousand in the four-year period, and serious labour shortages were experienced. By 1962 only three thousand were employed, and in 1973 only two uranium mines remained open, employing a mere sixteen hundred. Recently the industry and town have experienced another boom, again as a result of government contracts – this time from Ontario Hydro. The "cyclonics" of mining are still with us, and the people blown about the most are the wage labourers employed – or unemployed – by the industry.

Capital's relationship with labour is ambiguous. On the one hand, human labour is the source of capital's surplus and thus crucial to the realization of profits; on the other hand, capital strives to minimize the amount of "living" labour it requires by introducing "stored-up" labour in capital-intensive technology. The reasons for the latter are twofold. First, this strategy reduces the cost of production and hence improves profitability, and second, it minimizes the amount of direct control the capitalist must exercise over labour. Both reasons are central to understanding the dynamics of capitalism. Capitalists are required to accumulate or die. They must expand their operations through profit realization or be absorbed by other capitalists. At the same time, capitalists wage a continuing struggle with labour. Labour must continually strive to gain sufficient wages to sustain itself and, particularly in Canada, it must continually struggle to find a place where it can sell its labour power. These two processes are the central dynamic of capitalism. Only by placing them at the centre of analysis is it possible to understand, explain, and change the course of Canadian history.

CHAPTER THREE

An Anatomy of Inco Limited

The mining and processing of nickel required large amounts of capital, even more than was required for mining in the mountainous Kootenay region, and access to advanced technology. After a brief period of entrepreneurial capitalism, the nickel-copper industry in Canada was soon rationalized and consolidated. The history of Inco Limited, the giant of the industry, is one of mergers, acquisitions, and giant projects designed to maintain its dominant international position and with that its profitability. In this chapter we will briefly review the history of Inco as a corporation and examine its major locations in Canada and abroad.

The nickel-copper area of Ontario that was to become Sudbury was discovered as a source of minerals in the mid 1880s when the Canadian Pacific Railway was blasting its way through the Canadian Shield on its way west. It was copper that attracted the first entrepreneurs. In 1885 Samuel J. Ritchie formed the Canadian Copper Company to mine the area's outcroppings of copper. A smelter was built at Copper Cliff in 1888 and the Murray Mine, the first in the area, was opened in 1890.

Earlier, in 1877, the Orford Copper Company (the other half of the roots of Inco) had begun to smelt copper and sulphur in the Eastern Townships of Quebec but encountered opposition when sulphur fumes from its plant destroyed the neighbouring farms. As a result, the company had moved to New Jersey, where it was producing sulphuric acid for sale to the Standard Oil Company. The Canadian Copper Company used Orford Copper to refine its copper matte,* but it was found that the copper was contaminated by nickel.

Following the discovery and commercial production of nickel-steel in France by Henri Marbeau in 1885, nickel was transformed from a nuisance into a valued mineral, particularly for making armour-plate steel. The U.S. Navy, encouraged by the promotional efforts of Ritchie, became interested in the possibilities of this alloy and placed orders that provided the new Canadian industry with its initial market.[1] After some experimentation, by 1892 the Orford process for extracting nickel was developed, and in 1902 the Orford Copper Company, headed by Robert M. Thompson, and the Canadian Copper Company merged in New Jersey to form the International Nickel Company of Canada. Canadian Copper had owned the best ore bodies in Sudbury and accounted for 71 per cent of the nearly two million tons of nickel mined in Canada between 1887 and 1902; the Orford Company had developed a process and facilities necessary to refine the metal, thus making the new company the world's leading nickel producer from the outset.

International Nickel soon attracted the attention of powerful U.S. financiers and was taken over by J. P. Morgan and Company, which already controlled the United States Steel Corporation. As O. W. Main points out, "The new company, International Nickel, came under the financial control of [J. P.] Morgan and the new administrative control of the United States Steel Corporation. The first president, Ambrose Monell, was formerly with Carnegie Steel, and five of the directors were United States Steel officers."[2] The reasons for Morgan's interest were obvious. International Nickel could supply his steel company with a captive source of nickel, which had become a strategic resource for the armour-plate industry, thus assuring domination of the market; moreover, with Morgan's financial power he could profit from promoting the stock of the new company. International Nickel was, until the

* Matte is the product of the smelting process; it contains high quantities of metal along with sulphur and other impurities, requiring refining before pure metals are obtained (see p. 172).

early 1970s, controlled by U.S. financial interests associated with the house of J. P. Morgan. Today, control is in the hands of some of the world's leading financial organizations.[3]

In the summer of 1903, when International Nickel had been operating for only a year, there was a scarcity of labour, and the Sudbury mines had to be closed briefly so that all available workers could be utilized in construction. Work finally got under way on a smelter that could convert the ore to a Bessemer matte* containing 80 per cent nickel-copper. Further local refining was inhibited, however, by the U.S. tariff, and Ontario government attempts to force the company to refine nickel in Canada met with threats to develop its newly purchased ore bodies in New Caledonia. According to H. V. Nelles, "Every Canadian effort at challenging this tight monopoly was checkmated by International Nickel's political influence at Ottawa or control of the New York capital markets, or discouraged by the intractable qualities of the Sudbury ore, the capital-intensive nature of the refining industry, and the American tariff."[4] International Nickel continued to do its refining in New Jersey in the company's old Orford Copper plant, refusing to build a Canadian refinery.

Between 1914 and 1918 wartime demand for the "wonder metal" was phenomenal. The Ontario government, under the pressure of war, established a commission to investigate the possibility of nickel refining in Canada. As a result of the Ontario inquiry and the work of the Munitions Resources Commission, the federal government informed International Nickel that it must construct a refinery in Canada because of the importance of its product for the Empire's needs.[5]

Threatened with nationalization, International Nickel chose Port Colborne, Ontario, as the location for a Canadian nickel refinery, close to the U.S. market and to supplies of hydroelectric power. Construction of the Port Colborne electrolytic refinery began in 1916 and was completed in July 1918, just before the war ended. The nickel industry, which had enjoyed strong wartime markets, immediately began to feel the price of peace. The company laid off the workers and closed the Port

* Bessemer matte is the product of the bessemerising process whereby impurities are removed from molten metal by blowing air into it during smelting.

Colborne refinery in the autumn of 1921, not to reopen until the spring of 1922.

Among International Nickel's competitors was Mond Nickel, a British company founded in 1900 and enjoying lucrative contracts from the British government. In Sudbury it was working the Frood Mine in the same ore body as International Nickel's sources. From 1902 to 1914, Mond and Inco enjoyed a common market agreement, but it was disrupted by the war. The difficulties were resolved through an issuing of International stock to Mond, and the two companies were merged in 1929 under the International Nickel name. Under pressure of anti-trust action in the United States as a result of the impending merger with Mond, in 1928 the International Nickel Company of Canada was made the parent company of the organization, and shares were exchanged with the International Nickel Company (New Jersey), which became the subsidiary for the U.S. market. It continued to sell mill products from its Huntington, West Virginia, rolling mill. The merger with Mond brought into the organization another set of operations, particularly a nickel refinery at Clydach, Wales, a rolling mill in Birmingham, England, and a smelter in Coniston, near Sudbury; it also expanded International's market considerably in Europe. Although Canada had gained a refining capacity for the nickel industry, the actual manufacturing of nickel products still took place outside the country, leaving important manufacturing jobs to Inco's foreign operations. Another enterprise of 1929 was the formation of the Ontario Refining Company, with International Nickel and the American Metal Company each owning 42 per cent. In 1935 it became a wholly owned subsidiary of International Nickel and is now the Copper Cliff Copper Refinery. By 1930 the company had a 90 per cent share of the world's nickel market and retained some 80 per cent until the outset of the Second World War.

Falconbridge Nickel, Inco's Sudbury colleague, was formed in 1928 and benefitted from the high prices that resulted from the International Nickel monopoly. Between 1950 and 1957 the U.S. defence department stockpiled $789 millions worth of nickel, incidentally giving a boost to producers other than Inco, particularly Falconbridge, that brought them into competition with Inco. Adding to the strength of the other companies,

Falconbridge among them, was the development of laterite ores in tropical countries, a move not countered by Inco until the early 1970s. Falconbridge was developed under the guidance of Thayer Lindsley, an engineer and financial wizard.[6] After an initial period of Canadian ownership, control passed to various continental financial circles, finally coming to rest with Howard B. Keck (owner of Superior Oil in Texas) with the gracious assistance of the Canadian Imperial Bank of Commerce, which held the crucial balance of shares.

The next major development associated with International Nickel occurred in the mid 1950s when nickel was discovered in northern Manitoba. The search for these deposits began in 1946 using aerial geophysical prospecting that cost $10 million up to the first discovery in 1956. Another $15 million were spent on further exploration over the next five years. During the same period $140 million more were spent to open and develop the Thompson operation.[7] In his official history of the company, the former chairman of Inco, John F. Thompson, says of the board of directors' deliberations over the total $200-million expenditure for exploration and development of Thompson, "the directors continued to put up this money in the belief that they were morally obligated to the shareholders to protect the ore reserves the company already had, and to provide other reserves for the future."[8] Thompson, Manitoba, is the child of corporate capitalism at its most advanced stage, including generous state subsidies in the form of road construction and hydro-electric facilities. The mines and plants were constructed full-scale out of the wilderness and equipped from the outset with the most advanced mining and processing technology. The third-largest city in Manitoba is totally the product of decisions taken by Inco's board. And since the development of Thompson, the board has directed an international expansion of the company into Indonesia and Guatemala and a conglomerate-like diversification into other activities.

Inco from the beginning enjoyed a near monopoly in nickel, having merged with and absorbed many of its major international competitors. Since the Second World War its *international* market control has been weakened (but not broken), although inside Canada it continues as the single dominant nickel company. Its point of reference, however, is not Canadian but international. As Inco's monopoly of nickel was being threatened the pressure increased to expand both in Canada and abroad. There was also a demand for use of more effective production techniques, lest the new competitors produce the same product at lower cost. This possibility has contributed to the increase in capitalization to be discussed in detail later.

The capitalists in control of Inco have been among the most aggressive and opportunistic in the world. Since the turn of the century they have systematically taken advantage of wars; they have polluted the environment and resisted pressure to clean up or prevent pollution; they have seduced governments into granting great concessions and subsidies; they have often dealt savagely with their labour force by closing down operations and imposing layoffs; they have diversified and internationalized their operations, shifting capital generated in Canada to other areas. Part of the company's power has come from its financial ties to the leading edge of international capitalism, and part of its immunity has come from the fact that it does not deal directly with individual consumers. Instead, its customers are other giant corporations with whom Inco has developed close alliances, so that it would not be possible, for instance, to organize a boycott against the company. Throughout its history the executives of the company have been driven by one object, profitability, and have been particularly successful in achieving it, especially during wartime.

The Canadian Operations of Inco

The mines and surface plants collectively known as Sudbury are dispersed throughout the region. Except for the "superstack" that towers above the town proper, it is necessary to go to Copper Cliff to see the plants. Everyone knows that Inco is the dominant industry in the community and the basis of the area's economy. The Inco plant in Port Colborne is hidden away from the core of the town, across the Welland Canal and back from the visibility of the commercial district. What the townspeople see, right along the canal, is the skeleton of the abandoned Algoma Steel works. Now that Inco operations have also been reduced to a mere shadow, the townspeople who do have work commute to the automobile factories located in St. Catharines or other outlying plants, and the town is a bedroom community. Thompson, Manitoba, *is* Inco. The plants and mines are to the south of the town, clearly visible to anyone arriving by air or road. The fate of the town is identified with the fate of Inco. Everything else supports the mines. Sheridan Park, located between Hamilton and Toronto, is a modern research facility. Inco and a number of other companies house their researchers in this pleasant community. Aside from the Toronto headquarters, which has only a few employees, these are the sites of the major operations of Inco in Canada. A brief history and overview of each one will give some context for a more detailed examination of the actual workings of the plants and mines.

Sudbury

Before 1875, when most mining in Canada was for gold, silver, and coal and there was little systematic extraction of ore, the mining communities were made up of temporary bunkhouses, owned and operated by the mining companies, with little permanent settlement and few families. As the railways opened a number of areas and new minerals were discovered, towns such as Thetford Mines, Quebec, Kimberley, British Columbia, Glace Bay, Nova Scotia, and Timmins, Ontario, were established in the decades at the turn of the century. Sudbury was one of these, but it was not initially a mining town in the usual sense; rather, it acted as a service centre for smaller communities at the outlying mine and surface sites. Its population of one thousand at its incorporation in 1893 grew to eighteen thousand by 1931. Meanwhile, "most of the early settlements at the mines disappeared after 1900 as the mining companies centralized their smelting operations and built the permanent company towns of Copper Cliff, Creighton, Coniston, and Levack."[9] Not until 1920 did Sudbury itself become the home of nickel miners; about 25 per cent of its 1931 labour force and 50 per cent of its 1941 labour force were engaged in mining.[10]

The vegetation throughout the area was destroyed during the early years by the fumes given off in the process of reducing the ore. In the original "roast heap" method, huge piles of logs were covered with ore and ignited; they smouldered for months on end as the sulphur in the ore was burned off. At the turn of the century there were over eighty roaring heaps burning in the Sudbury area. Later, when the smelter was built and released sulphur into the air, an even wider area was affected. As an evasion of the pollution problem, "smoke easement" clauses were written into land sales around Sudbury after 1915, removing the right of the property owner to sue for damage attributable to sulphur emissions. More recently the air in the Sudbury area has improved, although acres of defoliated, blackened rock remain; Inco has built the huge superstack that disperses its pollutants over a wider area, thus reducing the immediate impact on Sudbury.

Sudbury remains Canada's largest company town. Its fate continues to rest on the extraction of nickel-copper, and its fortunes rise and fall according to the policies of Inco officials. The boom-bust cycle has not gone away. When the mines are producing, as in 1971, Sudbury may have Canada's lowest housing vacancy rate, but when layoffs occur, as in 1973, the rates can climb to the highest in the country.[11]

* * * * *

A long-time Sudbury resident (miner and union official): *Inco controlled the whole community, including Sudbury. They had a police force. It must have been in 1969 that we exposed it. When you had an inquest, the police who would investigate it were Inco police – Copper Cliff police, as they were called. Inco owned the cars. They used to buy those police cars and sell them to the police for one dollar. You could go down Copper Cliff Highway one day and get a speeding ticket from the policeman, and the next day you would find him at the gate of Inco checking your badge. . . . If there was an accident or a fatality and an inquest was called, the police would have to go and investigate it; it would be the same police. One day he's looking after the security at Inco gates; the next day he's giving you a speeding ticket; the next day he's investigating an accident.*

* * * * *

The people of Sudbury have always found it difficult to diversify the city's economic base because they have had little success attracting secondary industry. There are several reasons. One is the high cost of transportation to the major population centres; another is that other industries would have to compete with the high wages of nickel workers and combat the high cost of living in the area. A further consideration is the militancy of the workers and the fear that workers in new industries would be influenced by the area's unions.

At the turn of the century, the ethnic composition of Sudbury was primarily British (56 per cent in 1901), about one-third French Canadian, and only 10 per cent from other ethnic groups. In the outlying settlements near the mines, the other ethnic groups were more numerous.[12] In both Sudbury proper and the surrounding metropolitan area, persons of British origin continue to make up the largest group (37 per cent in 1971) but are now nearly equalled by French Canadians (32 per cent in Sudbury and 37 per cent in the metropolitan area). The remaining one-quarter to one-third of the population is composed largely of Italian, Ukrainian, German, and Polish ethnic groups, so that the area is one of the most evenly balanced in the country in terms of ethnic composition.

Reflecting the wages of the mining industry, incomes in the Sudbury area tend to be well above the national average (average weekly earnings in Sudbury were $199.44 compared to $178.08 for Canada in 1974), although the cost of living in the community is also high. Men receive the higher incomes in overwhelming numbers, for few women work in the mining industry. At the end of 1974, for example, Inco in Sudbury employed only 282 women (1.7 per cent of its labour force) and Falconbridge only 100 women (2.2 per cent).[13] The higher male to female ratio (1.11 to 1) in the twenty- to sixty-four-year-old age group persists because when women growing up in the area find few opportunities for employment, they respond by moving out.

Port Colborne

Here, more dramatically than in any of Inco's other operations, the effects of the weak nickel market (especially for Class I nickel – Inco's main product) are felt. But the current state of the market is not Port Colborne's problem; production and employment have been dropping for some time. The effects were evident in the early 1970s when the number of employees was reduced by layoffs from 2,350 in September 1971 to 1,779 in March 1972. As of July, 1977, there were 1,130 hourly, 199 staff, and 172 other workers employed in the research station for a total of 1,501 Port Colborne employees. Even before the most recent layoffs, only about 50 per cent of the operation's capacity was in use, and major sections had been shut down.

On September 8, 1977, Inco announced that it was cutting back its Port Colborne work force by another 384 hourly workers. By October, 97 of these workers said they would accept voluntary retirement pensions and 28 had quit, leaving 259 to be laid off. The number of hourly paid workers was now reduced to about 750. After this latest cutback, Port Colborne will still produce nickel rounds and utility shot,* but the plant has become basically a stock-shed for Inco and will be primarily a shipping depot.

* * * * *

A Port Colborne union official:
I have never seen the city of Port Colborne in a worse position than today [summer, 1977]. *Algoma* [Steel] *is shut down; Canada Cement is shut down; we are going through attrition. Right now Port Colborne is a dying community. You are never going to see this plant rolling again. Anyone who believes that is only dreaming. . . If it came to the worst, all you would see here would be four or five hundred people because they made it the central shipping depot.*

* * * * *

Inco workers at Port Colborne were hoping their town would receive the new nickel pellet plant, but it was instead built at the Copper Cliff Nickel Refinery which would reduce the ship-

* Nickel rounds and utility shot are specialty types of nickel, differing only in form from cut electrolytic nickel or nickel pellets as produced in the carbonyl process, produced because some customers find them easier to handle. See pp. 188-90.

ping costs of the raw materials. The town of Port Colborne has been heavily dependent on Inco. Its other major employer, Algoma Steel, began closing its operation, and the only other industrial employers were two flour mills and a shoe factory, which had closed down late in 1972. Early in 1977 some one hundred employees of the Canadian Furnace Division of Algoma were laid off, completing a long-term slowing down of the company's operation in the Port Colborne area. A joint Manpower Adjustment Committee combining the resources of Algoma Steel, Local 1177 of the United Steelworkers of America, and Canada Manpower was established to look into the problem. At its dissolution, after some six months' work, the committee reported that "of the 89 employees who had originally requested assistance to find employment, 63 were now employed or were receiving a pension."[14]

Of major significance is the final observation of the committee: "Perhaps the most significant factor of the entire exercise was the heavy demand for skilled labour, including trades embodying machinists, millwrights, electricians, pipefitters, welders, and maintenance mechanics. Placements could have been made for many times the number available. . . . Conversely, it was most difficult to place persons with limited knowledge of the respective trades." It is apparent which laid-off employees will be left on the unemployment rolls.

As we have seen, the selection of Port Colborne as a site in 1916 was precipitated by government action against Inco, forcing it to do more of its processing in Canada. The original advantages were the proximity of the U.S. market and the supply of power from Niagara Falls. Until recently, it remained the only Inco nickel refinery in Canada; now it has been surpassed by the refinery in Thompson, Manitoba, and the Copper Cliff Nickel Refinery. In the words of a company official at Port Colborne, the nickel refinery there is "about as significant as a pimple on a bull's ass." Indicative of the general condition of Port Colborne is the fact that only six apprentices have been hired since 1972, when the last big layoff came.

Today the Port Colborne plant's capacity is rated at about 200 million pounds of electrolytic nickel a year. During the early 1940s it was producing 250 million a year, dropping to a low of 160 million pounds in 1946 and rising to 216 million in

1947 during an arms build-up by the United States. Thereafter it has followed a downward trend, 160 million in 1951 (breaking 200 only between 1955 and 1957, 1959, and 1961), 164 in 1965, 84 in 1977, and now it has virtually been taken out of production.

Thompson

Much more than any other Canadian site of Inco's operations, including Sudbury, which has attained some level of maturity, Thompson is under the command of the company. This northern community is subject to severe weather conditions. Of its 130 annual frost-free days, only 30 are continuous, and it is a region of discontinuous permafrost. Winters in Thompson last an incredible 205 days; spring lasts only 23 days, summer 70 days, autumn 53 days, and the remaining two weeks are "between seasons" (everyone – justifiably – complaining about the weather). Isolated from Winnipeg by 450 miles, this is an unlikely location for Manitoba's third-largest city, but it is there, completely through Inco's initiative.

Unlike the traditional pick and shovel prospecting of earlier mining sites, an airborne search covering 150,000 line-miles of surveying with sophisticated equipment led to the discovery of Thompson. Additional ground searching covered 11,000 miles. Finally diamond drills intercepted a rich nickel ore body in 1956. At the end, twenty diamond drills were used to map out the deposit near Moak Lake. There had been no surface indications that such an ore body existed because it was buried beneath thick muskeg and swamp. Property rights were immediately staked by Inco. Its six thousand mineral claims covered a block eighty by five to six miles and cost the company $1.8 million. The initial labour force consisted of 150 men working drills, 75 to 100 geologists, and 30 to 40 native people clearing timber for transportation and power lines.*

* See E. S. Moore, *American Influence in Canadian Mining* (Toronto: University of Toronto Press, 1941), p. 104; Inco "Discovery" Sheet (Thompson, n.d.); John F. Thompson and Norman Beasley, *For the Years to Come: A Story of International Nickel of Canada* (New York: G. P. Putnam's Sons, 1960), p. 275; Hedlin Menzies and Associates, *Thompson, Manitoba: An Historical Impact Analysis of Resource Development* (Winnipeg: Hedlin Menzies and Associates, June 1970). The Menzies consulting firm, commissioned by Inco for the impact study, has done several other studies for min-

Development of the site posed many problems. Because the nearest railway connection was thirty miles east, construction materials were moved overland by tractor train. Operation Snowball, as it was called, was begun in January, 1957, to take advantage of the frozen ground. An around-the-clock stream of "trains" composed of three diesel caterpillar tractors and sleighs took about fourteen hours to complete a round trip. By April of that year, 30,000 tons of material consisting mainly of lumber, cement, and fuel oil had been transported. At its peak, construction employed over two thousand workers, and the major work was done by the end of 1959. Besides the surface operations and mines, a townsite had to be cleared and built. Initially 450 acres were cleared to accommodate a town of eight thousand inhabitants. Construction on this site was particularly arduous. In places, eighteen feet of permafrost were removed to reach bedrock for foundations.

Following the first year of construction a spur line was built from Thompson to the existing railway. This was necessary not only to get supplies into Thompson but to get nickel out. Of the $5-million cost of this Canadian National Railways line, Inco paid $250,000. Equally critical to the Thompson operation was access to a large supply of electricity. Manitoba Hydro obliged by building the Kelsey Generating Plant between 1956 and 1962 at a cost of $53.8 million, toward which Inco provided a $20-million loan, repayable in instalments up to 1980. Its sole customer was Inco (and the town of Thompson). In addition, the federal government spent $7.3 million on roads and an airport and the provincial government $7.5 million for roads and $3.6 million for schools. Of the total cost of the townsite, amounting to some $64 million, Inco contributed about $11 million. In conjunction with the provincial government, the company appointed a resident administrator to oversee the town's creation. In March, 1961, the town was officially opened and production of nickel began; in five years

---

ing interests: *The Impact of the Report of the Royal Commission on Taxation with Respect to the Development of New Mines in Canada* (1967) for Brenda Mines Ltd.; *The Impact of Coal Mining Operations of Kaiser Resources Limited on the Canadian Economy* (1968) for Kaiser Resources Ltd.; *An Analysis of the Kierans Report on Mining Policy* (1973) for the Mining Association of Manitoba.

Thompson was created out of the bush. Inco took the initiative and various levels of government provided the financial incentive.

From the outset housing has been a major problem in Thompson.

* * * * *

An Inco employee who came to Thompson in 1961 and lived at the company's campsite for two years before finding a private residence:

There were tents and bunkhouses. You stayed out in the tent first when you came. There was only a certain amount of room in the bunkhouses. The tents would sort of overflow, and as you stayed here longer, then you could apply to get out of the tents or they would just pretty well move you out of the tent because it was assumed that you didn't want to live there; so they moved you to the bunkhouse. I was there for two years, until the strike in 1964. During the strike they shut down all the bunkhouses and they were re-opened after the strike.

If they closed the bunkhouses during the strike, where did the men who were living there go?

They mostly all left town. That strike was particularly devastating to the community. In 1964 we started out with two thousand people working in the plant, and when we came to pay strike assistance three weeks later, we found we had only five hundred to pay.

When you were in the tent, how was it heated?

It had a space heater. They are pretty warm, really. They use them in the bush here all the time. Drilling outfits use them. I'll tell you, it's not a tent in the ordinary sense of the word. It's got ribs, framed in wood and covered with canvas, a small raised wooden floor, and a space heater. You are liable to have it too hot rather than too cold, which leads to colds and illnesses.

What was it like in the bunkhouse?

In the bunkhouse we had double rooms, but they were not completely up-dated. The walls would go up and then there would be a space for heating. It was easier to heat that way. And so you would have a space over the top, and you were constantly under surveillance of the security guards. They were really like a law unto themselves in those days. The security guards enforced the traditions in the camp, and they had official policies which were sort of at variance with the Criminal Code, I suppose. They set up a special building for gambling and things like that; you were forbidden to have alcohol in the camp, but as long as you weren't rowdy, they would turn a blind eye to it. But if you were rowdy, they would crack down on you. In the early days they were very, very anti-union; later on they came to accept the union, and the union was always pretty aggressive then. The workers were pretty young; at one time the average age was twenty-three. There were a lot of riots in the camp where people just rose up; I don't care how good the food is, if you get it out of the same kitchen over a period of time where you just have a limited amount of selection – say five different types of meals – let's face it. . . And the cooks got paid pretty low wages, too.

* * * * *

The company had a small residential community in Thompson from 1959 forward, but initially it was small (most men living in the tents or bunkhouses) and after a few years Inco sold its interests in housing. A worker who wanted to bring his family had trouble with the cost of living as well as with finding a place to stay. One miner who came to Thompson in 1962 stayed in the tents for two and a half months before finding an apartment and then had to pay $125 plus utilities, for a total of $150 monthly, while he cleared only $162 for two weeks' work, working seven days a week. Food and lodging in the tents cost eighty dollars a month in 1964 and buying a house cost a minimum of $140 a month in payments; heating bills often amounted to another seventy-five dollars a month. By 1968 the situation was little improved. It still took up to six months to find accommodation for a family, and fifteen hundred men were still living in a trailer park. Private individuals went into the price-gouging business. Rent on a two-room basement apartment with one bedroom and a shared bathroom was

$125 a month, and the starting wage was only $2.72 an hour.

Even today the housing situation is difficult. A family of four needs over $15,000 in annual income just to stay even and more if they want to buy a home. Table 1 illustrates the preponderance of rented housing in the town and the relative importance of apartment living. Since the recent cutbacks in hiring, the housing market in Thompson has again fluctuated. This time there is a surplus. In December 1977 there were 96 single-family dwellings for sale; by February 1978 there were 128, while the number of mobile homes for sale increased from twenty-one to fifty-one over the same three months.[15]

As would be expected, given the living and working conditions in Thompson, building a labour force is a perpetual problem. The turnover rates are incredible. The original construction crew for the town came largely from Manitoba because much of the work was done in the winter when farmers were available. The company provided food and lodging, and labourers were paid about $400 a month, semi-skilled workers $600, and skilled carpenters and electricians about $900. The work was long and difficult and living conditions were trying. Once operations got under way, most of the workers were employed in the mines and plants of Inco. Their numbers rose from 2,555 in 1962 to 3,538 in 1968. At the same time, until the late 1960s, there was a large pool of about 1,600 working on residential and commercial construction in the town. Once most of the building was complete this sector dropped off. During the early years it was thought that there would be about 2,350 Inco employees at Thompson; but this projection was soon raised to 4,025, and the total population of the town was expected to be about 12,000. In fact, by 1975 the company was employing 3,600 people and the town had grown to 20,000. There is no question, even with the growth of service industries, that the town remains dependent on Inco. As one miner put it: "We have got one place where we can work and that's Inco in this town; and if Inco doesn't work, we don't work. Nobody works. The whole town doesn't work."

The structure of the labour force is such that there are few jobs for women in the community. In 1963 women made up only 8.6 per cent of Thompson's labour force, increasing slightly to 10.9 per cent in 1966. By 1970 there were twelve

Table 1

Housing in Thompson, Manitoba (1976)

	Single houses	Attached houses	Apartments	Mobile homes	Total dwellings
Owned	1,480	80	85	260	1,905
Rented	680	260	1,880	35	2,855
Total	2,160	340	1,965	295	4,760

SOURCE: Adapted from Energy, Mines and Resources Canada, *Mining Communities* (Ottawa: 1976), p. 17.

Table 2

Age of Inco's Hourly Rated Labour Force in Thompson
(percentages)

Age	1963	1966	1969	1972	1975	1977
18–23	40	35	48	34	34	23
24–29	34	33	28	32	31	31
30–35	17	18	15	17	18	22
35 and over	10	14	10	17	18	24
Total	101	100	101	100	101	100
Number (year-end)	2,433	1,590	3,283	2,910	2,903	2,249

men working for every woman working; in the fifteen to forty-four years age group there were two men for every woman in the town and among singles 4.3 men for each woman.

While the over-thirty-five age group has shown a relative increase over the years, most of the labour force remains under thirty years of age, as Table 2 illustrates. The youthfulness of the labour force because of high turnover and the lack of employment for women are two main characteristics of Thompson. A third is the fluctuation in the number of people in the labour force as the labour requirements of the company rise and fall.

The town remains for many of its inhabitants a temporary stop. The transitory work force is made up mainly of young, single men who are not completely accepted by the "permanent" residents.

* * * * *

A seventeen-year resident of Thompson:
There's a great deal of discrimination against people who

live in the Polaris [single men's quarters]. *Real or imagined, but the discrimination does exist, I think generally because people tend to consider them as being here for a short time. If there is a great deal of turnover, it's in Polaris. If there is anybody going to leave, more likely it's from the Polaris than any place else. So, generally, those people have no roots in the community.*

* * * * *

Although the population of Thompson is 20,000 it seems much smaller. Everyone lives within the boundaries of the city, which is not a service centre for any hinterland. There is really no central core, the town being built in a circle. Most of the buildings are of steel siding and give the community a temporary feeling. The original shopping plaza has floors that seem to undulate as you walk on them, but the new plaza and the government buildings, particularly the Manitoba Building and City Hall, are solidly constructed.

Only the Union Centre has entertainment, and there are a few other taverns; but the government closed down two cabarets by revoking their liquor licences. Many residents feel that the government's licensing standards in Thompson are different from those in Winnipeg. People who like living in Thompson tend to take up fishing, water recreation, and winter sports. The trees around the town, mainly jack pines, are scrawny, the tallest being about twenty feet, and you can fit your hands around them. The permanent residents say they enjoy the size of the community and the neighbourliness. The union hall is an important community centre, and the union is active in food and gas co-ops as people attempt to make their lives more enjoyable and struggle against the high cost of living.

As will be discussed later, there is a lot of conflict in the town between the company and the union, but there is also some feeling of a common plight. As one union officer said, "This division is, in my opinion, very much a branch plant operation of Inco. Major policy decisions and attitudes are made in Toronto and Sudbury and not here. Some of the management personnel aren't operating their own store."

Sheridan Park

The J. Roy Gordon Research Laboratory is located in the Sheridan Park Research Community between Toronto and Hamilton. It is Inco's principal laboratory for research on extractive metallurgy. The company owns 9.4 acres at the main research facility and an adjoining fourteen acres. The laboratory itself is valued at $4 million and employs about 130 people: forty-eight professionals, sixty-six technicians, and twenty support personnel.

The Sheridan Park complex was initiated by the Ontario Research Foundation in 1963, and Inco, along with Dunlop Rubber and Cominco, were the founding participants. The community is designed primarily as a pleasant setting for "high calibre professionals" in the oil, rubber, pulp and paper, mining, and other industries. It has an association that draws these people together for seminars, and they help each other with technical problems.

The Inco laboratory is divided into four basic research sections: mineral processing, hydrometallurgy, pyrometallurgy, and electrochemistry. There is also a small geological research group. Of the research done here, about one-third is for improving existing milling, smelting, and refining procedures, another third for developing new processes for these activities, one-fifth for new ventures, and one-tenth for basic research. This is not Inco's only research facility. There are two larger laboratories in the United Kingdom and the United States. These deal with nickel products such as various alloys and are closely aligned with Inco production operations in these countries. Their field is marketing, while Sheridan Park is concerned with the actual processes for extracting nickel.

One major activity of the research facility has been the development of extraction processes for the lateritic nickel ores* of Indonesia and Guatemala. It is also active in research on nickel from nodules from the ocean floor. Within a consortium, Inco is involved in mining these nodules. It has taken responsibility for the processing end while one partner is looking after shipbuilding and another the mining technology.

* Laterite ores are of an oxide type in which nickel is concentrated just below the earth's surface as a result of leaching; they are found in tropical regions.

Inco's Internationalization

Since the merger with Mond in 1929, Inco has been a multinational company, but its multinational activities were for many years confined to the processing, refining, and fabrication of nickel products. Recently the company has expanded its extraction of nickel to other countries and has begun to diversify its activities beyond the nickel industry. The international operations of Inco have to be examined briefly in order to show the broader context of Inco as a corporation. They have had and will continue to have an important impact on Inco's Canadian operations. Inco's interests now extend to Guatemala, Indonesia, the United Kingdom, and the United States, with some operations in Australia, the Philippines, Mexico, Brazil, Papua-New Guinea, and Africa.

Guatemala and Indonesia

Inco made its ventures into Guatemala and Indonesia to protect its position as the world's largest producer of nickel, to which the laterite ores of these countries posed a potential threat. It began development in Guatemala in 1968 and in Indonesia in 1973. It assumed that world demand for nickel would grow to meet this new capacity, and in any event, besides preventing competitors from developing these ores, the new investment would weaken the bargaining power of its Canadian employees.

The projects that Inco has created in Guatemala (through its subsidiary Exmibal) and Indonesia (through its subsidiary P.T. International Nickel Indonesia) are for extraction and smelting of ore. The plant in Guatemala, completed in 1977 at a cost of $235 million, has an annual capacity of 28 million pounds of nickel. The much larger Indonesian operation has a potential capacity of 100 million pounds of nickel and cost $850 million. It is expected to be completed in 1979, and the first stages are in operation, shipping 8 million pounds of nickel by the end of 1978. In recent years over half the company's capital expenditures (two-thirds in 1977) have gone into developing the laterite nickel in these two countries.

Both projects produce a matte containing about 75 per cent nickel. Since the laterite nickel is the result of leaching of the soil, it is located near the surface and can be extracted by strip mining. The Guatemalan mines use huge power shovels and thirty-five-ton trucks to haul the ore. The cost of labour in that country is very low, but because of the mechanized extraction techniques only nine hundred people are employed, even though this is Central America's largest private capital investment. The matte from this project will be shipped to Clydach, Wales, for refining. The matte from the Indonesian project, in the first phase at least, will be shipped to Japan where it will be refined by some of the minority shareholders in Inco's project. It is expected that these Japanese steel companies will take about fifty million pounds or nearly half the total capacity. The Indonesian project will probably employ about thirty-five hundred workers when complete, but labour is not a major consideration where wages are only thirty cents an hour.

Inco unsuccessfully applied in 1969 and 1973 for exemption from Ontario's regulation restricting export of nickel matte (even though Falconbridge has been permitted to ship its nickel matte to Norway for years). Inco wished to export this low-processed product to Japan while the Ontario government wanted refining to take place within the province. Now Inco is able to supply this market using Indonesian laterite ores as its source.

Inco has had little difficulty raising the capital for the South Asian project. Its own equity capital accounts for about a third of the funding. The other two-thirds was procured through long-term debt. Jamie Swift gave this outline of the financing:

> The first phase syndicate was managed by the Bank of Montreal, while the larger of the second phase consortia was managed by the Citicorp International Bank of London. The latter is owned by Citicorp, a holding company for the second largest bank in the United States, the First National City Bank. Included in this syndicate were the Bank of Montreal, the Toronto-Dominion, the BNS International (Hong Kong) – a unit of the Bank of Nova Scotia – Morgan Guaranty Trust, Crocker National Bank, Chemical Bank of New York, Bankers' Trust Company, and the Asia Pacific Capital Corporation.[16]

Thus North America's most powerful financial capitalists have a major share in financing this giant project. A small portion of the shares are held by the Japanese steel companies, and there is a provision that about one-fifth of the shares will eventually be sold to Indonesians. In the long run, however, Inco will hold about three-quarters of the shares.

In a backhanded way the Canadian people have also "invested" in the undertaking. Aside from the fact that the company's own profits over the years have come from its extraction of ore in Canada, the Canadian government has provided Inco with several additional sources of capital. Inco has received $378 million in tax deferral benefits, which amount to interest-free loans; in 1977 it was given a $10-million grant for reserve depreciation; and the Export Development Corporation has lent the company, at low interest, $74 million to expand nickel production. In 1973 the EDC made a loan of $17.25 million and in 1976 another of $40 million for the Indonesian project. The government's attitude is, "If we had not lent this money somebody else would," according to Acting Prime Minister John Munro.[17] The main beneficiaries of the EDC loans, aside from Inco, are two equipment and service companies in Canada, Canadian Bechtel and Montreal Engineering, both U.S. subsidiaries.[18]

When the Indonesian and Guatemalan projects came into production just as Inco was implementing a cutback of some four thousand jobs in Canada, a furore naturally arose. There was little likelihood, however, that Inco would cut back the other projects. The first five to ten years of such enterprises enjoy substantial tax-holidays that would make it unprofitable to slow down production. Moreover, the company felt that if it were to impose cutbacks in these nations, there would be repercussions. Walter Curlook, a senior Inco vice-president, said, "I don't think there's any doubt in our minds that Third World countries like Indonesia and Guatemala are much more likely [than Canadians] to act quickly against Inco if we took measures that would seriously affect their social and economic development programs."[19] The hold these countries have over Inco should not be exaggerated, however. In both projects the end-product is matte. In itself this product is saleable on the world market only to the very few companies with refining capacities for nickel, and these companies, like

Inco, have multinational operations of their own to protect. Moreover, there is always the threat of military intervention in the event of nationalization. The United States has made interventions for less, and both Indonesia and Guatemala are dependent upon U.S. hegemony in the current world political, economic, and military situation.

The tropical laterite projects do not have all the advantages when compared with the sulphide ores of Canada. The smelting of laterites requires a great deal of energy, much of which comes from petroleum. Recent increases in the cost of petroleum have made it necessary for Inco to continue to raise the price of nickel, even in the face of weak markets.* As a result the sulphide ores (containing sulphur, which in essence burns itself) can compete better with the laterites, even though they cost more to remove from the ground and Canadian wage costs are much higher. In addition, laterite ores contain much less nickel than sulphide ores per ton and lack the copper and precious metal content of the ore in the Sudbury basin. Yet there is a paradoxical twist resulting from the energy situation. Because of the high cost of petroleum, Inco added a hydro-electric power plant unit to its Indonesian project, scheduled for completion at the end of 1978. In order to make the electricity project economically viable, it was necessary to expand the capacity of the Indonesian project threefold.

It is apparent that the development of these laterite ores expands Inco's capacity greatly and helps return the company to the dominant position it held from the turn of the century until the Second World War. Inco has reacted against other mining companies that were eating into its monopoly position and at the same time has protected itself against strikes in Canada. It is no longer possible for Canadian workers to "bring Inco to its knees" as they did in 1969 when the market was strong. Inco could continue to supply its customers from its laterite projects and the refinery in Clydach.

Inco's United Kingdom and United States Operations

Inco's refinery in Clydach, Wales, the largest in Western

* According to a report by J. P. Schade, a senior vice-president of Inco Metals, "each price rise of $1 a barrel [of petroleum] adds 5 cents (U.S.) to the cost of producing a pound of nickel from a good-grade lateritic ore" (*Globe and Mail*, 17 Nov. 1979: B6.)

Europe, has a capacity of 45,000 tons of nickel a year and refines about a quarter of the company's nickel. It is a carbonyl refinery* producing nickel powders and pellets, so that it can provide both Class I and Class II nickel. According to the *Annual Report for 1976*, by 1978 the company intended to add a "fluid bed roasting plant and an associated sulphuric acid plant to treat the roaster gas. The new facilities will give the refinery the ability to treat nickel matte produced in Guatemala and Indonesia."[20] The company also has the world's largest precious metals refinery in Acton, England, that produces platinum, palladium, rhodium, ruthenium, and iridium, all precious metals that the company ships unrefined from Canada in the form of a valuable black sludge. Another British subsidiary, Henry Wiggin and Company, produces cupronickel, high-nickel alloys, wrought nickel, and other nonferrous nickel alloys. In 1975 Inco purchased Daniel Doncaster and Sons of Sheffield, England, a metals forging and machinery company.

Since the United States is the major market for refined nickel, Inco does most of its product development there. Its main plant is the West Virginia mill, where it produces cold-drawn rod, cold-rolled strip, highly polished sheet, wire, and tubing, and high-grade nickel alloys. These are all semi-fabricated shapes and nickel products, none of which are produced in Canada. It also has a formed metal plant in Burnaugh, Kentucky. The research laboratory for developing new nickel alloys and products is also located in the United States in Wilmington, North Carolina. In 1975, through its subsidiary Inmetco, Inco invested $30 million to open a production plant in Ellwood City, Pennsylvania, for converting the wastes from specialty steel mills. Another Inco project in the United States is the development of uranium deposits, through its subsidiary American Copper and Nickel Company, in a joint venture with the Tennessee Valley Authority on some three hundred square miles of land owned by the Ford Motor Company in upper Michigan.

* Carbonyl refining of nickel was developed by Ludwig Mond and Carl Langer in 1889. Nickel carbonyl is a colourless gas which, when heated to 180°C. (365°F.), separates into carbon monoxide and pure nickel. It first came into production in 1895 in England using matte produced in Canada.

Inco's Diversification

Besides moving into multinational mining activities Inco has also embarked on a diversification program into areas outside the nickel industry. Most of them have nothing in common with nickel but profitability. The smallest is farming and cattle ranching in Guatemala on lands the company controls; the largest is the purchase of ESB of Philadelphia, the world's largest battery manufacturer, at a cost of $241 million. ESB produces Ray-o-Vac batteries for digital watches, maintenance-free car batteries, halogen lamps for public places, and long-life transistor radio batteries. It operates ninety-six plants, nine located in Canada and others in South Africa, Brazil, Guatemala, and fourteen other countries, and employs 17,000. There have also been investment projects, such as the $500,000 spent to gain a 10 per cent interest in United Tire and Rubber Company. Inco has a 4.5 per cent interest in Panarctic Oils Limited and a half-share interest in a uranium mine along with Canadian Occidental Petroleum Limited in northern Saskatchewan.

These diversifications are now a major part of Inco's operation, precipitating the change of its name in 1976 from the International Nickel Company of Canada. About 50 per cent of the company's revenues are now derived from business other than primary metals, although 70 per cent of its assets remain in that field. These managerial strategies to diversify are certainly not ones Canadian workers would have made, nor is it likely that they would have bought Daniel Doncaster and Sons or Diado Special Alloys Limited, a joint-venture in Japan. They would probably have created this productive capacity in Canada.

The final expansion Inco is undertaking may, in the future, be the most important. Along with several other multinational companies from Germany, Japan, and the United States, Inco has a major share in Ocean Management Incorporated and Sedco Incorporated. Both are engaged in developing sea-bed mining. At the sea's bottom there are potato-shaped manganese nodules that contain high quantities of nickel, copper, and cobalt. The plan is to dredge up the nodules into a ship with a ten-inch diameter pipe three miles long (see Figure

Fig. 1. Sea-bed mining

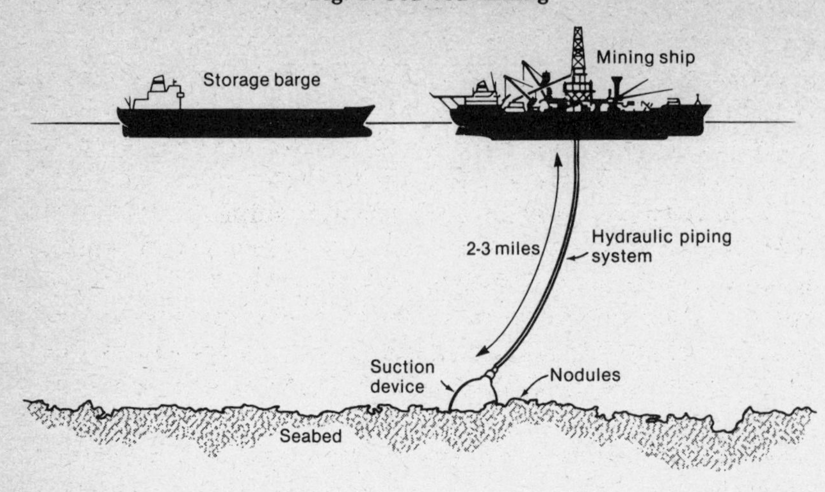

1) by means of a vacuum and then transfer them to freighters. The first full-scale test is scheduled for 1982. The main impediment to this undertaking, apart from the technology, is ownership of rights to these minerals, so that Inco has a very large interest in Law of the Sea conferences designed to establish these rights. It is attempting, through representations to the U.S. government, to ensure that its interests are well supported. For the United States the sea-bed source provides an opportunity to supply "domestically" some crucial minerals, thus reducing its external dependence.

As a corporation Inco has experienced important changes in the past decade. It has stepped up its internationalization and diversification programs in an effort to stabilize its profitability. It has also engaged in important reorganization within Canada, both on the surface and underground. The various mining operations of Inco differ considerably from one another. Technical considerations and the nature of the ore play an important role in the significance of each site. The ore from laterite mines contains between 1.25 and 1.5 per cent nickel, about the same as the ocean nodules, which in addition contain about 35 per cent manganese. Laterite ores contain about 0.1 per cent cobalt but no copper. In contrast, the sulphide ores of Canada have a higher content of nickel and, particularly at Sudbury, large quantities of copper and precious metals.* The relative value of each ore depends upon the price that nickel, copper, or the various precious metals command. In addition, there are significant processing costs to be considered. According to an Inco researcher, "the big difference between laterite ores and the sulphide ores is that with the sulphide ores you can take a low grade material, grind it, go through flotation, and up-grade it by a factor of ten.... Very economic in terms of tonnages, and you very quickly reduce the tonnage costs down to a reasonable level. In the oxide ores [laterites] and sea nodules you have to process a whole lot. There might be some up-grading where you take out the boulders and the six-inch rock, but the ore's all mixed in with the rock. You can do a screening, a rough up-grading, but it might reject only 25 per cent of the rock instead of 80 per cent." These off-shore ores therefore cost more for transportation, processing, and energy. Not only do the Canadian sulphide ores contain sulphur and burn themselves but Canadian smelters also have access to large quantities of relatively cheap hydro-electric power.

Political considerations are also important. Although the company feels secure in its Canadian operations, workers in Canada have been successful in attaining good wages and have proven willing to fight for their rights, safety measures, and even pollution controls. In Third World countries the political situation is less secure, but wages are low and lucrative agreements with governments have been struck. There is little pressure concerning environmental issues and, even more important, if Inco does not develop these deposits, other multinationals may.

* In 1972 the Inco mines in Sudbury produced 155,000 short tons of nickel and 159,000 of copper while the Thompson mines produced 63,000 short tons of nickel and only 4,000 of copper.

CHAPTER FOUR

Markets, Labour, and the Mining Industry in Canada

Just how important is the mining industry to Canada? Measured against the total economy in 1973, it accounted for only 1.0 per cent of employment and 1.7 per cent of wages and salaries. In terms of capital expenditures it was somewhat more impressive, accounting for 2.8 per cent. Its real significance in terms of the economy was shown in its 4.3 per cent of the value of production and particularly its 10.4 per cent of the value of merchandise exports.[1] Mining has its primary economic importance as an export sector and is used by the state to attempt to offset trade imbalances caused by imports of manufactured products. Besides the crude mineral exports that account for 10.4 per cent of total exports, there were also major exports of fabricated minerals, which means that over 20 per cent of all exports from Canada consist of mineral products.[2] Over 70 per cent of Canada's mineral production is destined for the export market, much of it in a crude form with little value added, while Canada imports large amounts of fabricated mineral products (much of it in the form of equipment used in mining).

Canada continues to rely on a small number of minerals as the basis for this trade. In 1974–75 only three minerals – iron, nickel, and copper – accounted for 45 per cent of crude mineral exports and together with the next four – zinc, coal, asbestos, and potash – accounted for over 80 per cent.[3] The principal destination of these exports is the United States. The

Table 3

Canadian Minerals and the United States *(1970–73 average)*

Commodity	Canadian exports to U.S. as % of Canadian production	Imports from Canada as % of U.S. imports	Imports from Canada as % of U.S. consumption
Asbestos	41	97	87
Nickel	46	63	57
Potash	71	95	57
Gypsum	74	77	29
Zinc	34	55	24
Iron Ore	46	51	16
Silver	71	52	52
Sulphur	23	72	9
Lead	24	31	16
Copper	21	37	7

SOURCE: Adapted from Energy, Mines and Resources Canada, *Mineral Industry Trends and Economic Opportunities* (Ottawa: 1976), p. 21.

situation of nickel is not unusual in Canada's trade relationships with the United States, as Table 3 indicates. All the top ten minerals exhibit similar characteristics: Canada is the major source of the raw materials, and these raw materials, in turn, represent a significant part of U.S. consumption. The vulnerability of Canadian resources to U.S. market conditions is plainly illustrated.

Canada's mining industry grew dramatically as a result of U.S. demand for base metals associated with the Second World War and the Korean War. The Paley Report (*Resources for Freedom*), commissioned by the U.S. president and issued in 1952, identified twenty-two key resources necessary to maintain the United States as the world's leading industrial nation. Canada was designated as the major source for twelve of them. The report was only officially sanctioning a trend that had already started.

The share of Canadian mining and smelting controlled by foreign capitalists rose from 38 per cent in 1946 to 57 per cent in 1953, continuing to 70 per cent by 1957. In the course of a decade, Canada's mining industry ceased to be Canadian and became foreign dominated. The U.S. government provided corporations with foreign investment incentives, as did the Canadian government. The 1955 Canadian federal budget made "permanent special tax concessions" to mining that included "a three year tax holiday for new mines, a depletion allowance amounting to 33 ⅓ per cent of net profits, deduction of all exploration and development costs ... and a special capital cost allowance on mine buildings, machinery and equipment."[4] It should be noted that these incentives did not foster Canada's processing of minerals any more than they improved Canadian ownership. Indeed, by 1974–75 a greater share of mineral exports were in crude form (54 per cent) than had been the case as early as 1928–29 (40 per cent) or 1955–56 (44 per cent) – particularly for copper, nickel, and zinc.[5] Extraction and mineral manufacture each accounted for half the total capital formation in the mineral industry in the early 1950s, but by 1971 mineral manufacturing was responsible for only one-third and extraction for two-thirds of the industry's capital formation.[6] The clear tendency, shown in Table 4, has been to put increasing reliance on cruder natural resource exports with less and less processing in Canada before export. These changes result mainly from two factors.

Table 4

Proportion of Selected Minerals Refined in Canada, 1950 and 1970
(percentages)

	1950	1970
Copper	90	67
Lead	100	53
Zinc	65	38
Nickel	54	50
Silver	84	69
Iron Ore	38	18

SOURCE: Adapted from P. E. Nickel et al., *Economic Impacts and Linkages of the Canadian Mining Industry* (Kingston: Queen's University, Centre for Resource Studies, 1978), p. 89, Table 24.

One is that the United States, the European Economic Community, and Japan place higher tariffs on mineral imports as they move into more advanced processing, thus encouraging importation of minerals with little processing. The other is that most of the extraction is done by foreign-controlled corporations and is designed to feed natural resources to their manufacturing operations at home. Although some of these corporations use Canada to take advantage of Commonwealth preferential tariffs, most are restricted by their parent companies from exporting Canadian manufactured products. Canadian General Electric, for example, a major fabricator of copper products, is not allowed to export these products from Canada.[7] Even though Canada is the world's largest exporter of nickel, asbestos, zinc, and silver and second in potash, uranium, molybdenum, and sulphur, its position has not been exploited as a source of leverage in dealing with advanced capitalist nations. This is largely because most of the firms exploiting these resources to supply their own raw material requirements are controlled outside Canada.

In addition, the strong concentration in the metal mining industry gives great power to a few giant firms. Table 5 gives an indication of the relative concentration by type of metal. Even

Table 5

Concentration in Metal Mining Production in Canada, 1972
(percentages)

Largest corporations	Copper	Nickel	Lead	Zinc	Molybdenum	Gold	Silver
One	24	79	45	30	91	22	26
Two	43	96	73	53		41	48
Three	53		90	75		57	56
Four	59			85		65	62
Five	65			89		70	68

SOURCE: Adapted from Elizabeth Urquhart, *The Canadian Nonferrous Metals Industry: An Industrial Organization Study* (Kingston: Queen's University, Centre for Resources Studies, 1978), p. 42, Table 9.

this understates the actual concentration because Inco, for example, is not only first in nickel and second in copper but also prominent in silver and gold.

The metal mining industry in Canada is characterized throughout by high concentration, export orientation, low degree of processing, capital-intensive operations, and foreign control. In each of these characteristics the nickel industry is typical of the metal mining industry as a whole.

Markets, Production, and Prices of Nickel

Nickel is one of the key resources necessary for industrialization. It is essential for making stainless steel, structural steel, and machine parts. Of worldwide production, 40 per cent is used for stainless steel, 16 per cent for electroplating, 14 per cent for high-nickel alloys, 11 per cent for construction alloys, and 9 per cent for iron and steel castings. It is the backbone of a good deal of primary and secondary manufacturing, and demand for its qualities increases with more sophisticated technology. For example, although "a four engine piston driven plane required only about 125 pounds of nickel . . . the Boeing 747 uses 11,000 pounds. . . ."[8] Two-thirds of the demand for Inco's nickel stems from capital goods spending and only one-third from consumer goods.

Inco's proven reserves of ore in Canada in 1976 were 415 million short tons, valued at about $35 billion at current prices, containing about 13.4 billion pounds of nickel (sufficient for about thirty years' demand), and 8.6 billion pounds of copper. These are only proven reserves that have been sampled and calculated; there is probably much more.[9] While reserves of nickel are substantial, it is a non-renewable resource and is defined as ore only as long as it can be processed at a profit. If other sources of nickel, such as the laterite ores of the sub-tropics or the sea-bed nodules, become more economical to extract and process, Canada's ore will turn into rock.

As can be seen in Figure 2, world nickel shipments have increased impressively, if somewhat irregularly, since the Second World War. Recently the deliveries of nickel have swung with the health of the economy. Until 1969 there was a fairly steady increase, but a major strike at Inco in that year caught the company with low inventories and a higher demand than could be met from stockpiles. The strike effectively cut Inco's 1969 production by 50 per cent, but in 1971 both production and demand increased. In 1974 demand was up from the previous year, but supply could not keep up. At the end of 1974 inventories were at their minimum; Inco's deliveries were up 20 per cent over the previous year, and production reached 500 million pounds. By year-end 1975, production was down 10 per cent from 1974, again mainly because of strikes, but deliveries were also down 25 per cent.[10] In 1976 Inco reduced production to 462 million pounds, and the 1977 figure was down to 415 million pounds, according to John McCreedy, chairman of Inco Metals. Deliveries that year dropped to 312 million pounds (the lowest since 1958) with a production capacity of 600 million pounds. By the end of 1977 inventories stood at 341 million pounds, compared with the usual 100 million.[11] Consequently, massive layoffs took place – a common occurrence in this cyclical industry. Along with the layoffs, operations at Birchtree Mine in Thompson and Copper Cliff North Mine in Sudbury were suspended; near Sudbury, production stopped at Creighton No. 3 and was reduced at the Stobie section of Frood-Stobie, work at Crean Hill Mine was suspended in mid 1978, and the small Victoria Mine was closed permanently. The result of the layoff followed by the strike in September 1978 was to reduce inventory by 111 million pounds by the end of the year.[12] Each month of the strike reduced it a further 20 million pounds. At the beginning of 1979 there were signs of a renewed market and the possibility of nickel shortages by the end of the year.*[13] Hence the boom-and-bust fluctuations continued as year-end sales for 1978 increased to 377 million pounds.[14] At the end of March, 1979, Inco's nickel inventory totalled 146 million pounds. By the strike's end in June, 1979, the inventory was well below the normal 100-million-pound reserve.

Fig. 2. World nickel shipments

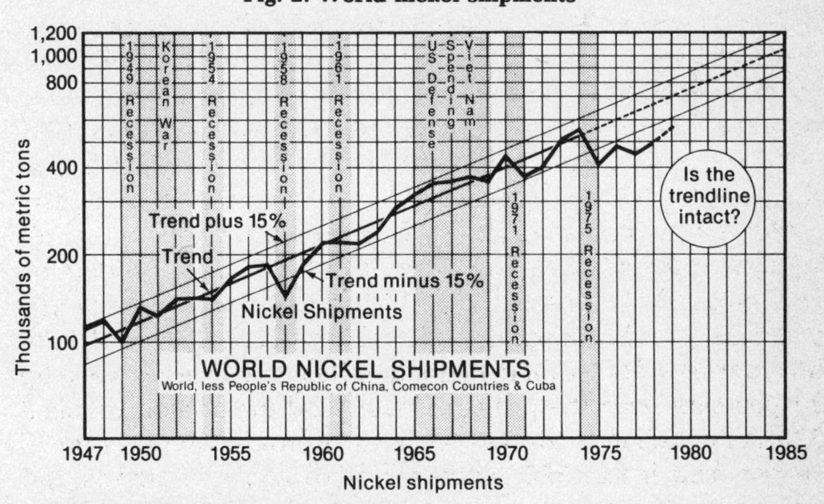

Reproduced courtesy of Falconbridge Canada

* Aside from inventory reductions as a result of the 8½ month strike in the Sudbury operations of Inco, world production was reduced substantially in March 1979 by an explosion and fire which were expected to close the main refinery of Société le Nickel, a French competitor, for ten months.

Whatever happens at Inco shakes the entire Canadian nickel industry. As Table 6 illustrates, Inco produces 85 per cent of Canada's nickel, and only Falconbridge has a significant share of the rest of the market. The effects of slumps in the nickel market are felt most heavily in Ontario and Manitoba. Before Thompson came on stream in 1961, virtually all of Canada's nickel production was located in the Sudbury basin and, as Table 7 indicates, three-quarters remains in Ontario.

Table 6

Canadian Nickel Production, 1975

	Metric tons, 000	Per cent
Inco Ltd.	208.5	85.2
Falconbridge Nickel	28.0	11.4
Noranda Mines	2.7	1.1
Sherritt Gordon Mines	2.6	1.1
Dumbarton Mines	2.6	1.1
Kanichee Mining	0.4	0.2
Total	244.8	100

SOURCE: Based on Energy, Mines and Resources Canada, *Nickel* (Ottawa: 1976), p. 31.

Table 7

Production of Nickel by Province, 1970 and 1974

	1970		*1974*	
	Short tons	Per cent	Short tons	Per cent
Ontario	244,255	73.3	234,473	78.2
Manitoba	79,121	25.9	64,456	21.5
British Columbia	1,704	0.6	732	0.2
Quebec	801	0.3	...	...
Canada	325,881	100	299,661	100

SOURCE: Based on Statistics Canada, *Canada's Mineral Production*, various issues; Energy, Mines and Resources Canada, *Canadian Minerals Yearbook, 1974.*

Needless to say, the booms and busts make a considerable difference to profitability. In spite of the cycles, Inco has not failed to turn a profit since 1932, although there have been large fluctuations. As Appendix II illustrates, Inco's return on investment sank from 21 to 13 per cent between 1974 and

1975. Its 1976 profits rose again to $196.8 million, but profit declined to $99.9 million in 1977 and further to $77.8 million in 1978 (all U.S. dollars). On the other hand, it is clear that share of the market makes an important difference. Falconbridge suffers even more from the cycles and over all has a substantially lower return on investment.

The effect of market fluctuations on nickel prices is most interesting. As the dominant producer of nickel in the world, Inco sets the price. It has continued to increase the price steadily to pay for its heavy investments in developments in Canada and the rest of the world, particularly since the mid 1960s, when demand for nickel was higher than production. Falconbridge follows Inco's lead. At the end of 1974 Falconbridge attempted to increase the price of nickel to $2.05 a pound but soon rolled back to Inco's increased price of $2.01. Despite the lack of demand for nickel, Inco again increased the price to $2.20 in August 1975, following its July contract with the United Steelworkers in Ontario; Falconbridge followed suit in September. In November 1976 Falconbridge tried to raise the price to $2.53, but when Inco announced its price of $2.41 a few weeks later, Falconbridge again backtracked. As Table 8 illustrates, the price of nickel has risen steeply, doubling in

Table 8

Price of Refined Electrolytic Nickel

Date	$U.S. per lb.	Date	$U.S. per lb.
1950	.51	1976	2.41
1955	.63	1977	2.36
1960	.70	1978	2.41
1965	.84	March 1979	2.30
1970	1.38	April 1979	2.55
1973	1.53	May 1979	2.90
1974	2.01	June 1979	3.05
1975	2.20	December 1979	3.25

the 1970s, falling back, then recovering during 1979. Inco was consistently able to increase its price, even during adverse market conditions, until April 1977, when it announced for the first time that it was reducing the price of nickel by five cents a pound. The reduction was a reaction to other producers that had been eating into Inco's monopoly position over the previous few years by offering discounts on the list price.

Then, in July 1977, Inco announced that its published prices no longer applied, that prices were now "confidential business information." According to John McCreedy of Inco Metals, "We simply decided that Inco wasn't fair game for target practice, that we wouldn't, couldn't take it lying down any longer."[15] Inco, having difficulty maintaining its monopoly position with weak markets and operating at such low capacity, had decided to strike back.

Not until nearly two years later, in February, 1979, did Inco again post its price. It had fallen from its 70 per cent command of the world market in the 1960s to a share of about 35 per cent.* Laterite ores from New Caledonia had gained a prominent share of the market. As noted earlier, Inco had responded with its own laterite projects in Guatemala and Indonesia. The rising cost of processing the energy-demanding laterite ores, however, has again strengthened Inco's Canadian operation, which derives its nickel from less energy-demanding sulphide ores. Inco may be able to recover some of its lost ground for this reason, but the company still has heavy capital debts. There is little danger that Inco will fail as a result, but Falconbridge is most vulnerable under these circumstances, and it is entirely possible that its future in the nickel industry may be limited.

There is little that can be done in Canada about fluctuations in nickel demand. Less than 5 per cent of Canada's nickel production is consumed at home; the rest, obviously, is exported (aside, of course, from stockpiles). Exports and international conditions make the nickel market go. The fact that so much nickel is exported reflects Canada's weak industrial position internationally. But nickel can be exported in various stages of processing. Only a small amount of Canada's nickel exports are in the form of fabricated metal, as Table 9 shows. While Canada has not exported nickel ore since 1974, it is clear that a large and increasing part of the ore produced is exported in a very low stage of processing as concentrates, and as a result a good many jobs and a good deal of value added are also exported. Appendix III shows that matte is exported primarily to

* Inco nevertheless remains the "lowest cost nickel producer in the world," according to Ilmar Martens, a nickel analyst (*Financial Post*, 28 July 1979:4).

Table 9

Canada's Nickel Exports by Degree of Processing
(percentages)

	1928–29	1955–56	1964–65	1974–75
Ores and concentrates	34.7	38.4	46.7	52.5
Matte, oxides, anodes, ingots, and rods	65.3	61.6	51.8	43.4
Fabricated items	...	...	1.5	4.0
Total	100	100	100	100

SOURCE: Based on Bruce Wilkinson, "Trends in Canada's Mineral Trade," in *Working Paper No.* 9 (Kingston: Queen's University, Centre for Resource Studies, 1978), p. 33, Table 10.

Norway and the United Kingdom, with some going to Japan. About two-thirds of the nickel oxides and refined nickel go to the United States. The matte shipped to Falconbridge's refinery in Norway and Inco's plants in the United Kingdom is destined for the European market after it has been processed and further refined, often into fine metals. As illustrated earlier (Table 3), Canadian nickel accounted, on average between 1970 and 1973, for 63 per cent of the United States' imports and 57 per cent of their nickel consumption. From Canada's perspective, the United States receives 46 per cent of Canadian production.

Although nickel is Inco's principal product, the company also derives a good deal of revenue from other metals (see Table 10). In Canadian copper production, Inco at 19 per cent is behind only Noranda Mines at 24 per cent. Inco can sell all the copper it produces without difficulty. Another significant by-product of processing nickel is cobalt, which ranges between 1.7 million and 2 million pounds annually from Inco's Canadian operations. With the price of cobalt leaping from $6.40 a pound in December 1977 to $25.00 a pound in February 1979, Inco is aiming at expanding its production to 3 million pounds. As noted earlier, Inco's non-nickel metal revenue is largely the product of the Sudbury operation. The late 1970s slump in the nickel market was not extended to the other metals. Stockpiles of these were quickly exhausted during the 1978–79 layoffs and strike at Sudbury, and the company lost this production along with the reduction in its stockpile of nickel. There was a precarious balance between the

Table 10

Inco Metals Sales, 1977

	$000 U.S.	Percentage
Primary nickel	557,924	64
Refined copper	211,101	24
Precious metals	63,276	7
Other	37,292	5
Total	869,593	100

SOURCE: Based on Inco Ltd., *Annual Report for 1977* (Feb. 1978).

company's need to reduce its nickel stockpile in order to generate revenues to pay its debts and its loss of revenues from these other products.

The market for nickel since its inception followed closely the international military situation and, along with that, the capital goods (such as heavy machinery) and consumer goods industries (such as automobiles). Inco's first real boost came with the First World War, after which it substantially reduced production. The Second World War had an even greater effect on expanding Inco's capacity, particularly since it was a war of machines, and those machines consumed great quantities of nickel. During the Second War Inco even processed matte for Falconbridge when its refinery in Norway was in German hands. Between 1939 and 1945 Inco delivered 1.5 billion pounds of nickel, resulting in a $35-million expansion of the Canadian operations and a $3.5-million growth at Huntington, West Virginia. As a result of the post-war concern in the United States to ensure a supply of nickel, the government contracted for 120 million pounds of nickel from Inco over a five-year period to build up stocks. These contracts, worth over $80 million to Inco, between 1946 and 1955 acted as the major source of stimulation for constructing Thompson, Manitoba. The Royal Commission on Canada's Economic Prospects (1955) was prompted to comment that U.S. government contracts were the most significant factor in nickel production at the time.[16] In 1960, when Inco was again understocked, it bought 51.4 million pounds of nickel from the U.S. government stockpile to meet its sales.[17] (This stockpile weakening labour's position in negotiating with Inco.) U.S. government defence spending on the Vietnam War between 1965 and 1970 again represented a major outlet for Inco.

Overproduction is not caused simply by weak markets

because markets for nickel today are much larger than they were in the past. The problem is an overestimation of the markets and a rapid expansion of capacity induced by attempts to restrain competitors. Moreover, as will be seen later, there is a long lead time between investment decisions and the opening of a new plant. Current investments are made in response to decisions taken nearly a decade ago. The $1 billion Inco spent between 1967 and 1972 developing laterite mines in Indonesia and Guatemala greatly expanded the company's capacity and its debt. Although the specific debt structure of Inco is not known, L. Arthur English, manager of metals and mining for the Toronto-Dominion Bank, wrote that loans to mining companies expanded from $101 million in 1961 to $882 million in 1972, accelerating especially after the 1967 revisions to the Bank Act. He reported, regarding the six largest Canadian mining companies, that "in 1966 total debt for these companies as a percent of equity was 2.9% but by 1971 [it] had grown to 51.5%."[18] It becomes obvious that Inco and other nickel companies must continue to raise their price for nickel to pay for their capacity expansion and pay off their debts. At the same time they have gotten themselves into a position of carrying large inventories. At the beginning of 1977 Inco had a five- to six-month stockpile, compared with its usual two- to three-month inventories, and it started 1978 with about a year's inventory. This represented a great deal of tied-up capital for a company with outstanding debts.

Since the Second World War, the cyclical demand for nickel, affected by the Korean and Vietnam wars, brought peaks every four to five years: 1953, 1957, 1961, 1966, 1970, and 1974. Inco's production cycles are also influenced by the creation of stockpiles just prior to contract negotiation years. This is its way of protecting itself whenever possible from threats of strikes. When stockpiling is not possible, as in 1969, the company finds itself vulnerable to workers' demands. Use of this strategy also undermines, to some extent, the notion that Inco is a "victim" of the stockpiles. Certainly the 1978 stockpile was greater than normal, but the company's strategy was to continue adding to it through high production in order to strengthen its bargaining position with the union for that year's contract.

From the outset the market for Canadian nickel has been international, developed by foreign capital to satisfy foreign industrial demands. Nickel has *never* been perceived by capitalists or the state as the basis for an indigenous Canadian industry. The state's maximum aim has been the primary refining of nickel in Canada, which still means exporting an unmanufactured raw material. Even this has not been achieved, and indications are that ground is actually being lost. Both the fundamental structural weakness of Canada's industry and the poverty of state policies are reflected in this situation.

Mining and Its Labour Force

The proportion of people employed in the mining industry as a whole, compared with the total labour force, has experienced significant declines, dropping from 1.3 per cent of the total Canadian employment in 1961 to 1.1 per cent in 1971. On an average rate over the decade, mining employment increased absolutely by only 1 per cent a year compared with an overall employment expansion of 3 per cent a year. Confirming the demise of independent producers in mining discussed in Chapter 2, there were only 1,235 self-employed workers in Canadian mining in 1971 contrasted with 137,650 wage earners. Table 11 shows how the number of persons employed in the nickel-copper industry fluctuated between 1963 and 1974.*

Labour has always been a cause of dilemmas in mining. In boom periods it is difficult to obtain an adequate supply, particularly in remote single-industry communities like Thompson and to a lesser extent in such places as Sudbury. The availability of skilled labour is uncertain. The work force is militant, for miners have traditionally been highly organized and tough, independent workers. Capital has responded to these labour problems in mining by introducing mechanization and automation and organizing this technology in such a way that workers are easier to train and the number of workers re-

* Obviously the layoffs at Inco and Falconbridge would bring more recent figures even lower. The impact of layoffs is highly concentrated; well over half of all nickel-copper workers are located in the Sudbury basin, with northern Manitoba the only other significant location.

Table 11

Employment in Nickel-Copper Mining, 1963 to 1974

Year	Number employed	Annual change
1963	10,425	
1964	11,546	1,121
1965	13,016	1,470
1966	13,225	209
1967	13,699	474
1968	14,980	1,281
1969	12,425	−2,555
1970	16,691	4,266
1971	17,664	973
1972	15,310	−2,354
1973	14,696	−614
1974	15,083	387

SOURCE: Based on Statistics Canada, *General Review of the Mineral Industry*, various issues.

quired is reduced. Thus we will see, particularly in the late 1960s and the 1970s, the rapid infusion of a new generation of technology into mining. The industry is currently in a stage of transition, but the directions are clear: use as much equipment and as little labour as possible.

Since Inco (like most other mining companies) is not involved extensively in marketing or fabricating products in Canada, an area where employment tends to increase as the number of workers in extraction and production is decreasing, the workers displaced have not been replaced in either the production or the white-collar sectors. The trend is for fewer workers to produce even larger amounts of nickel.

Labour shortages have been a perennial problem in the mining industry, especially for skilled workers. They were particularly acute during the Second World War, when although Inco obtained recruits through the National Employment Service there were still shortages of up to 20 per cent of the normal roster. The *Canadian Mining Journal* described how Inco sent out employment agents to hire workers throughout the country: "Farmers from the prairie regions were recruited for work during the winter months, and commercial fishermen from the Maritimes were recruited in the seasons when their work was slack. Construction workers, lumbermen, local farmers, and soldiers on short furloughs – in fact, everyone who could give assistance for either short or long periods."[19] As recently as 1974 the Mining Association of Canada argued that labour shortages and turnovers were costing its members over $350 million a year.[20] In 1973 labour shortages in mining caused a loss of production of about 8 per cent.[21]

A different type of labour shortage is caused by strikes. It is evident from Table 12 that the mining industry is hit harder by strikes and lockouts than either the construction or the manufacturing industries, losing over twice as many man-days per wage earner as the other two sectors. The strikes clearly do have an effect on the ability of mining corporations to produce, particularly during periods when the market is strong and inventories are low.

Table 12

Time Loss Due to Strikes and Lockouts in 1975

Industry	No. of hourly wage earners	No. of strikes	Duration (man-days)	Man-days per wage earner
Mining	79,400	44	1,173,070	14.8
Construction	173,200	111	1,064,580	6.1
Manufacturing	917,500	480	5,264,710	6.1

SOURCE: Calculated from Energy, Mines and Resources Canada, *Mineral Industry Trends*, p. 27.

Labour shortages have led to development of a much more capital-intensive quality in the mining industry. As Table 13 illustrates, the ratio of both production and value added to wages increased substantially in the nickel-copper industry in the 1970s. In the mining industry as a whole, the post-war period has certainly been one of rapid capitalization. J. A. MacMillan and his associates, who have made the most comprehensive study of this question, found that "in the period 1948–73, the dollar value of Canadian mineral productions grew almost eightfold, while the number of persons employed increased only 25 per cent, from 106 thousand (representing 2.1 per cent of the Canadian labour force) to 133 thousand (1.4 per cent of the Canadian labour force)."[22] In primary mining they found that employment increased only 14 per cent while the value of mine output increased eightfold over the twenty-five-year period.[23]

Table 13

Capital Intensiveness of Nickel-Copper Mines, 1970 to 1974

	Wages of production workers	Value of production	Production: Wages	Value added	Value: Wages
	$000	$000		$000	
1970	149,303	1,039,040	6.96	634,644	4.25
1971	159,779	889,617	5.57	448,779	2.81
1972	146,519	865,464	5.91	521,000	3.56
1973	149,720	1,167,636	7.80	813,843	5.44
1974	148,335	1,480,204	9.98	1,043,377	7.03

SOURCE: Calculated from Statistics Canada, *Nickel-Copper Mines, 1974* (Ottawa: 1976), p. 5.

Paralleling the increasing capital intensity of mining is the change in the proportion of hourly to salaried workers. In Table 14 we see the far greater increase in the number of salaried workers over hourly workers in the metal mining industry between 1948 and 1973. Appendix IV shows in more

Table 14

Hourly and Salaried Workers in Metal Mining, 1948 and 1973

			Increase	
	1948	1973	Number	Per cent
Hourly	37,705	47,984	10,279	27.3
Salaried	4,185	18,150	13,965	333.7
Total	41,890	66,134	24,244	57.9

SOURCE: Based on J. A. MacMillan et al., *Human Resources in Canadian Mining* (Kingston: Queen's University, Centre for Resource Studies, 1977), p. 13.

detail that metal mines tend to have a higher proportion of non-production workers than do non-metal mines, reflecting their greater capital intensity. Table 15 examines the relationship between hourly and salaried workers somewhat differently. In 1948 only one worker in ten in metal mining was on salary; by 1973 the proportion increased to over one in four. In terms of the actual salaries and wages paid, the increases are even more significant. Salaried workers received 13 per cent of the income total in 1948 but 33.5 per cent in 1973. Even as recently as 1970 production workers received 76 per cent of the total, but by 1974 this had dropped to 63 per cent.

Table 15

Employment in Canadian Metal Mining, 1948 to 1973

	Per cent production	Per cent salaried	Total
1948	90.0	10.0	41,890
1953	88.7	11.3	51,711
1958	85.3	14.7	61,999
1963	81.0	19.0	57,119
1968	77.7	22.3	63,369
1973	72.6	27.4	66,134

SOURCE: Calculated from Statistics Canada, *General Review of the Mineral Industry*, various issues.

It is possible to determine an increasing ratio of salaried to hourly workers, but ascertaining specific occupational changes in the metal mining industry is made difficult by changes in census classifications. Table 16 shows as nearly as possible changes between the last two census years. There have been noticeable increases in the proportions of workers in this industry employed in managerial, professional, and clerical jobs, all of whom would be salaried workers. The increase in transport operation indicates the greater mechanization in moving the ore while the proportion of people actual-

Table 16

Occupational Distribution in Metal Mines, 1961 and 1971

1961	Number	Per cent	1971	Number	Per cent
Managerial	1,279	1.9	Managerial	1,630	2.4
Professional	4,207	6.1	Science & engineering	5,570	8.2
Clerical	3,142	4.6	Clerical	4,340	6.4
Service	1,953	2.8	Service	1,395	2.1
Transport	1,517	2.2	Transport operation	3,135	4.6
Other primary*	37,091	53.8	Mining	28,080	41.3
Craftsmen	17,629	25.6	Processing	3,735	5.5
Labourers	1,580	2.3	Machining	2,495	3.7
Other	533	0.8	Construction	4,385	6.4
			Material handling	2,735	4.0
			Product fabricating	6,655	9.8
			Other	3,895	5.7
Total	69,931	100		68,050	100

SOURCE: MacMillan et al., *Human Resources*, pp. 122-24; see also App. V.

* Nearly all are miners.

ly engaged in mining has decreased. Only these fairly broad changes in the whole industry can be drawn from census data. The 1971 census, however, does provide a more specific indication of the actual occupations in metal mines. Appendix V sets out a finer breakdown of all occupations performed by more than five hundred persons. It can be seen that the industry employs a large group of scientists and engineers (nearly four thousand) and about fifteen hundred additional support workers related to science and engineering. Relatively few people are employed in clerical jobs. In mining itself about 15 per cent are foremen, 19 per cent drillers, and 34 per cent in cutting, handling, and loading. The only other major occupation, employing over six thousand, is mechanics and maintenance.

Workers in metal mining tend to have somewhat higher wages than workers in manufacturing. Appendix VI shows details of the annual labour costs for both groups. Aside from the miners' higher basic pay, the real difference between the two is in the bonus system, which accounts for about 10 per cent of wages in mining but is not a usual method of payment in manufacturing. The bonus tends to increase the wages of miners considerably over those of people in manufacturing. Later it will be seen that only certain workers in mining, particularly production workers underground, are on the bonus, which averages about $250 a month or $3,000 a year. Their total incomes are thus considerably more than these figures would indicate, while those of workers on surface and in non-incentive mining jobs are closer to manufacturing wages.

Appendix VII gives the actual hourly rate in 1973 for various occupations in mining. Most striking is the similarity for all the occupations, both on surface and underground. Mine helpers aside, the difference among all other underground occupations is less than sixty cents an hour. On surface the maintenance workers have the highest wages, but the difference still is not very great except for labourers. When bonus is added to the wages of underground workers, their total income matches and even exceeds that of maintenance workers who are not on a bonus system.

When the educational attainments of workers in the mining and manufacturing industries are compared, the mine workers tend to have the lower standing. Recently, however, the proportion of workers with post-secondary training in mining has increased to match that in manufacturing, indicating the occupational trends noted earlier and the trend toward larger numbers of salaried workers. In Table 17 we see the comparative educational distribution between mining and manufacturing for 1951 and 1971. The educational level has gone up to some extent in the mining industry, but nearly two-fifths of miners in 1971 continued to have only public school education in comparison with a third of the workers in manufacturing and 27 per cent of the Canadian labour force over fifteen. Among men working as miners, the proportion with public school education only rises to 45 per cent. Lack of formal education still appears to be an important reason why people enter and remain in mining. It is a relatively high-paying job where someone with little formal education can learn the necessary skills on the job.

Table 17

Educational Attainment for Males* in Mining and Manufacturing, 1951 and 1971
(percentages)

	Mining		*Manufacturing*	
Grade level	1951	1971	1951	1971
Less than nine	63.7	38.9	53.8	33.0
Nine to thirteen	29.6	49.9	38.2	55.8
Over thirteen†	7.0	11.3	8.1	11.2
	100‡	100	100	100

SOURCE: MacMillan et al., *Human Resources*, p. 126.

* Only males are included because virtually the entire mining industry is male dominated.

† For 1951, the grade level used was twelve.

‡ Sums do not total 100.0 because of rounding errors.

Traditionally, mining has been an occupation where many immigrants to Canada have found employment. The Ontario Royal Commission on Nickel, in its *Report* in 1917, observed that in Sudbury "the labour in the district is principally foreign, probably not more than 25 per cent being Canadian or American. The more skilled workmen, such as foremen, mechanics, and carpenters, are Canadian or American. Underground, the drill runners and helpers are principally

Finns and Austrians; the trammers are generally Poles, Italians, Austrians, and Russians."[24] The census of 1911, 1921, and 1931 indicate that immigrants made up over half (about 52 per cent) of Canada's labour force in mining but only about one-third in all industries. Originally immigration was a source of skilled miners, but after the turn of the century more immigrants were used as unskilled labourers. Today immigration is declining in importance as a labour pool for mining. Indeed, it is less important in mining than in other industries. In metal mines alone 17.8 per cent of the labour force in 1971 was born outside Canada; for all industries the rate was 20.2 per cent. Table 18 compares immigrants employed in mines, quarries, and oil wells with those in all other industries in 1961 and 1971. Over the ten-year period, the proportion of non-Canadian-born workers dropped most among workers in mines, quarries, and oil wells compared with Canadian industry as a whole.

Table 18

Immigration and Canada's Industry, 1961 and 1971
(percentages)

	1961		*1971*	
Born	Mines, quarries & oil wells	All industry	Mines, quarries & oil wells	All industry
In Canada	78.1	78.3	84.6	79.8
Outside Canada	21.9	21.7	15.4	20.2
	100	100	100	100

SOURCE: Calculated from Energy, Mines and Resources Canada, *Mining and Manpower* (Ottawa, 1976), p. 12.

The first woman ever to enter an Inco mine did so in 1939; she was Queen Elizabeth, wife of King George VI. In 1959, her daughter Queen Elizabeth II visited the Frood Mine. But these women obviously did not come to work. In Ontario women are still prohibited by law from working underground or at the face of open-pit mines, although they are allowed to do either in Manitoba. Moreover, the exclusion of women from the mining industry is not limited to underground. As is evident in Table 19, very few women work in the metal mines industry; in 1971 they made up only 4.2 per cent of the labour force (2,825 out of 68,050). Recent figures suggest the proportion is now around 5 per cent, but they continue to account for only 0.3 per cent of production jobs in mining.

The few women in the mining industry tend to be located in clerical jobs, which make up 36 per cent of the occupations held by women; the science-related jobs, service occupations, and actual mining (mainly in open-pit mines) each have 5 per cent of the women employees. They are obviously confined mainly to office jobs, where in 1973 the pay averaged only about $2.50 an hour compared to the $4.00 to $5.00 an hour average in production.

* * * * *

A Thompson hourly paid woman:
I was out looking for a job and early in the morning I went down to Inco and filled out an application form. The receptionist said, "I'm sorry, but we just don't have any jobs." There were a couple of women sitting there, and I started talking to them, and while I was talking there were several men that came in. Two of the men had worked there before and came back. They were hired. But she had just said there were not any jobs, and there might not be any jobs available for at least two or three months, so I had to accept that at face value. Then some more men came in, and these were new hires – I mean people like myself who were putting application forms in – and they got hired on the spot. Just like that. So I stayed there a couple more hours, and the receptionist kept telling me to leave because there were no jobs. So I came down the next week, the next two weeks. I just kept coming down every day, and I would just sit there, and she kept saying, "There's no jobs." I said I had nothing better to do and just sat there. She didn't know what was going on, [but] I noticed this whole routine of men getting hired. Men were continually getting hired.

* * * * *

Table 19 illustrates that women are virtually absent from production jobs, and only a few find their way into sales and distribution. Even in administrative and office jobs, however, there are still very few women. Arguments that women cannot

Table 19

Sex of Workers in Nickel-Copper Industry, 1970 to 1974

	Production		*Administrative & office*		*Sales & distribution*	
	Males	Females	Males	Females	Males	Females
1970	16,780	4	3,338	431	117	33
1971	17,650	5	3,696	468	108	23
1972	15,341	0	3,326	468	150	29
1973	14,698	0	4,410	737	152	34
1974	15,083	42	4,562	858	159	35

SOURCE: Calculated from Statistics Canada, *Nickel-Copper Mines*, p. 5.

do the work because it is physically too demanding simply do not hold for administrative and office jobs; not only that, they do not hold for most production jobs. Women have been systematically excluded from mining, and there is little evidence of change. Some of the barriers women encounter will appear later in some of the interviews. As the ratios of men to women in Thompson and Sudbury, discussed earlier, illustrate, the fact that there are few young single women in these communities contributes to the high labour turnover. And there are few because the mining companies, typically the major employer in the area, have provided them with few employment outlets.

* * * * *

An Inco official interviewed at Port Colborne:
Is it correct that, aside from the war, there have never been women working in the Port Colborne plant?

That's right. During the war there was an Order-in-Council put through that permitted the hiring of women in heavy industry, and we needed workers. We had women operating the casting wheels, we had women making boxes, women on the sewing machines, women in the store house, counter girls. We had a couple of darn good machinists in the machine shop, running lathes and wood-working tools; they did a very good job for us. At the end of the war this Order-in-Council was rescinded and we had to let them go.

You were forced to, by the government?

That's right. We were forced to. Men were coming back from overseas; there was a shortage of jobs and there was no way they were going to let housewives hold down jobs that men needed.

* * * * *

The first hourly paid women employees in Inco's Thompson operation went to work in January, 1974. By October of that year there were over sixty women working in various surface jobs – including the Pipe Mine. Subsequent layoffs have reduced the number of women working in the Thompson operation because of their low seniority, but at least there is no longer a total ban. In the Sudbury operations there are a few hourly paid women working in most of the surface facilities, but they are still not very noticeable. Women are explicitly excluded from major parts of the Copper Cliff Nickel Refinery because of gas exposure, but in most areas the justification for exclusion is economic. As a senior Inco official put it, "It's not our policy not to have women. It's the problem of trying to find the money to put the facilities in." This does not explain, of course, why the facilities were not available long ago and is simply indicative of the sexist attitude which permeates the entire mining industry, including the male miners, government, and management.

* * * * *

A senior Inco official at the Iron Ore Recovery Plant, explaining why no women work there except in the offices: *Nothing to do with chemistry or anything. Just a lot to do with the physical facilities available. We don't have any washrooms for the operations end of it. To bring women into the plants we would have to put extra facilities in, and we don't have them now and I don't know whether we will get them in in the immediate future either, but we have never had women working full-time out in the plant.*

* * * * *

The Canadian mining industry continues to be plagued by dramatic fluctuations in the demand for labour, oscillating between labour shortages and layoffs. The general trend toward more capital-intensive operations, and consequently to more salaried and fewer hourly paid workers, is an attempt by mining company management to reduce and control its demand for labour. As a result there are important increases in the proportion of workers engaged in managerial, professional, and clerical jobs, but the vast majority of workers in this industry are still engaged directly in mining and processing. Because mining companies are part of the monopoly sector of industry, they have been able to pay their workers relatively high wages, and because these workers have been organized and militant, they have made substantial wage gains over the years. Mining companies have traditionally relied on immigrants as a source of labour, but this pool is now drying up. The current labour pool for production workers in the industry consists of poorly educated young Canadian-born men. The industry has not - except during wartime - called upon that other source of cheap labour (as the textile and food and beverages industries have done) - women. Women have not been employed even in administrative or sales jobs in mining to any great extent. This affects the entire character of mining communities. Most notably in Thompson, but also in Sudbury, there are few employment opportunities for women, and the few that do exist tend to be in the lowest-paid jobs. As a result these communities tend to be unstable; a large part of the labour force is made up of young men passing through the communities and either quitting after a short work period or being laid off because of low seniority.

CHAPTER FIVE

It's Another World: Inco's Underground Operations

Gradually Gerald got hold of everything. And then began the great reform. Expert engineers were introduced in every department. An enormous electric plant was installed, both for lighting and for haulage underground, and for power. The electricity was carried into every mine. New machinery was brought from America, such as the miners had never seen before, great iron men, as the cutting machines were called, and unusual appliances. The working of the pits was thoroughly changed, all the control was taken out of the hands of the miners, the butty system was abolished. Everything was run on the most accurate and delicate scientific method, educated and expert men were in control everywhere, the miners were reduced to mere mechanical instruments. They had to work hard, much harder than before, the work was terrible and heart-breaking in its mechanicalness.

But they submitted to it all. The joy went out of their lives, the hope seemed to perish as they became more and more mechanized. And yet they accepted the new conditions.

From *Women in Love* by D. H. Lawrence.

The point of mining is, of course, to get the ore from the stope (or working face of the mine) to the surface. The basic stages for removing the ore are drilling holes in veins of ore, blasting the ore with explosives, removing the ore from the face, and dropping it down to ore cars in which it is transported for preliminary crushing underground before being hauled to the surface. These steps are common to all of Inco's mines but, as will soon be evident, the way they are performed varies greatly in different types of mines. Besides production work, which is actually getting the ore out, a great deal of the effort necessary to removal of the ore is development work, that is, preparing areas for drifts (passageways), ore passes (chutes to move ore between levels), and service shafts. Various factors influence the way this work is organized: the decisions made by management about the kinds of equipment used, the levels of skills the miners bring to these tasks, and the types of inducements used by management to get miners to perform their tasks. Conditions underground differ radically from those on surface. Traditionally miners have had great autonomy in organizing their work. They have usually worked in small crews involved in an entire work cycle, either development or production, and direct supervision has been minimal. Recently, management decisions about types of equipment used, training of miners, and methods to get the workers to perform their tasks have radically changed the way work is organized in the mines.

Placer mining, with its owner-workers, today accounts for less than 1 per cent of Canada's mine production. The basic relationship in mining is now wage labour, but when it is underground, wage labour takes on a different form. Virtually all hourly paid workers in the mines who can directly affect the rate of production or development work are on a bonus system (or incentive, as the company likes to call it). Yet like so much else in mining, this system also is subject to important changes.

* * * * *

A Thompson cut-and-fill miner:
Mining is one of the most interesting jobs a working man can get. Frustrating sometimes, very frustrating. Work like hell. It is not one of those jobs where you say, "Geeze,

when will the end of shift come!" There is no question that the bonus is an incentive. But we work just like animals. People that are in other industries just don't realize how hard a miner works. But mining is far more interesting than an assembly-line job or a surface job in the smelter or refinery.

* * * * *

The Mines

The means of removing ore from the ore body varies greatly from one type of mine to another. The four basic types of mines in Inco's Canadian operations are traditional hand-work mines (Crean Hill and parts of Creighton No. 5 and Creighton No. 9 in Sudbury), captured equipment mines (Thompson Mine in Thompson and parts of Creighton No. 5 and Creighton No. 9 in Sudbury), ramp mines (Creighton No. 3 and Levack West in Sudbury), and open-pit mines (Pipe Mine in Thompson). Within these types some of the basic mining methods used are the square-set timber method, the cut-and-fill method, and the blast-hole method, and more than one method is sometimes used in one mine. In this chapter we shall examine change in various phases of the actual mining operation using these different methods within these types of mines.* But before we do so it is worthwhile to have a general introduction to some features of Inco's mines for a basic understanding of the differences in their settings and methods.†

Crean Hill Mine

Crean Hill is located in a rural area about twenty-five miles from Sudbury that was once a company town. Production began here in 1899. It continues to use "traditional" mining techniques. Its daily four thousand tons of ore are transported by CP Rail to the Clarabelle Mill in Sudbury. The general mine foreman of Crean Hill has long experience in the mine and

* The mines reviewed in this section are a sample (see Appendix I) of the basic types of mines at Inco.

† Mining, like so many industries, tends to have its own language and terminology. Wherever possible, brief explanations are offered of these unusual terms, and reference is made to the section where they are explained in more detail.

knows almost all the 330 employees by name; they call him by his first name. He is an easy-going man but works from 6 A.M. to 6 P.M. five days a week and is on call most of the rest of the time, including every fourth weekend. This mine has the best safety record in the company.

Supervisors at the mine work nine-hour days and are usually in charge of more than one level. Crean Hill normally works seven days a week with three shifts. There are about 150 men on the afternoon and day shifts and 30 on graveyard, which is only a supply and service shift with no production. The mine has a very low turnover rate; no one had been hired since the previous September when we were there in July, and only a few had left. Most of these miners have homesteads and come from the area. They often do a little farming on the side. They are the personification of the adage that farmers make the best miners because they are used to steady physical work. The policy at the mine is to rely on the surrounding community instead of hiring workers from Sudbury.

Except for the hoist man, a skilled and respected operator who works in a building set away from the mine head and office area, few people work on surface. Two other men who do so work at maintenance of the hoist. They are not allowed to talk with the hoist man while the hoist is in motion. He works in an enclosed area, watching dials and listening for bell signals from the cage tender. He operates two Norberg hoists: a single-deck cage that can hold forty-three men and a skip (or self-dumping bucket) that hauls ore or rock to surface. During August, 1977, there was a three-week shutdown for repairs to the hoist. The workers had to take their vacations then, and those with less than three weeks due them took a week off without pay or worked elsewhere; it was a mini-shutdown in anticipation of the major one the next summer.

The entrance to the cage is foggy and damp. The men, who work in staggered shifts, put their tags on an in-out board and wait on wooden benches in an area just off the entrance to be called for their levels. The mine itself is dark, damp, and muddy. After leaving the cage underground, the men go through a series of heavy steel doors installed for fire protection and direction of ventilation. These doors are common in all types of mines, except, of course, open-pit mines. The walls of the mine are plastered with safety reminders, another feature common in Inco's mines. Not far from the shaft on most production levels is a lunch room where the mine foreman has a rough desk that he uses as an office; there are also two rows of benches, a small refrigerator, a hot plate, and a sink. At the start of each shift the miners traditionally have a coffee break before they begin work, reassembling again as the shift ends to await the return cage.

Underground, the miners at Crean Hill tend to work alone or, more often, with one or two others in small work areas. One man working alone in the dark on a fan-drill (see page 132) will see his supervisor only once a shift and the other men on his level only at lunch and before and after work, when he is waiting for the cage. The drilling is done in a small drift and is very noisy, dirty, and wet. The miner follows drilling plans supplied to him by the engineering section. He may have bid* from a stope crew for the job because the bonus is good, averaging 40 per cent of his wages based on the number of feet drilled, but he has to work alone and in the dark.

* * * * *

A Sudbury miner:
When you are walking in a drift, it is black. Everything is black and collapsed. You have got this light, and it kind of hypnotizes you as you are walking down. The walls are very close, and you get a very sharp light. You can't see more than fifty feet ahead of you because everything is black after that; you can't see anything to the side of your head.

* * * * *

A slusher operator (see page 140) and his partner on a tram (see page 145) move the ore from the stope through an ore pass to a dump. The former sits at a two-lever machine directing a blade attached by two cables to pull the ore into a train car

* To bid is to request a job that has been posted by the company as available. The person with the most seniority who is qualified for the job will receive it. Frequently miners are trained for more highly qualified jobs than they currently perform and must wait until the higher-rated job becomes available before they can move up.

which is tended by his partner, and when the ore car is full he leads the train to the dump and blows his whistle to warn anyone ahead. These men can directly control the pace of their work and are on a shared bonus, averaging about 25 per cent. The slusher area is one of the few in the mine that has any light aside from the lamps on the miners' helmets.

In Crean Hill, driving a raise (the passage between two levels) is done manually. This is one of mining's most skilled jobs, but fewer and fewer workers have the necessary skills. Raise crew miners are among the most respected in the mines. Their work, contrasted with the use of highly mechanized raise borers, will be examined shortly (see pages 136-40).

This mine has very little diesel equipment and has remained much the same since the Second World War. During the most recent layoff in 1978, it was announced that production at Crean Hill would be halted until demand once again expanded. This is as would be expected, because the cost of labour (which is variable) far outweighs the capital investment (which is fixed) in this mine.

Creighton No. 9 Mine

The Creighton complex, the oldest Inco operation, was begun in 1901 as an open pit (see Figure 3). It is located ten to fifteen miles from Sudbury. No. 9 shaft can produce eight thousand tons of ore a day and feeds the Clarabelle Mill. The labour force is about 250 men – 90 in stopes, 18 on development, 50 on trams, and the rest maintenance workers. Thus it can double Crean Hill's output with eighty fewer men.

Figure 3 shows that this mine is directly connected underground with No. 5 shaft. The main problem here is with heat and ventilation. The mine is 7,137 feet deep (the deepest single shaft in the Western Hemisphere), and at the 7,000-foot level the temperature at the rock wall is 108°F. Future development work is planned for a 9,000-foot level where the rock wall temperature will rise to 135°F. (the deeper one goes in the mine, the hotter the temperature, which rises 1°F. every hundred feet). As the grade of the ore seems to improve with the depth, Inco is intent on going deeper. Plans are under way to sink a No. 11 shaft as a ventilation system, passing the air through the ice fields in Creighton No. 3 mine (see page 107) to

Fig. 3. Cross-section of the Creighton complex. No. 9 mine handles everything mined below the 2,100-foot level. No. 3 is mining blast-hole stopes in the hanging wall of the ore body and hoisting at No. 7 shaft. Outlined areas indicate ore bodies.

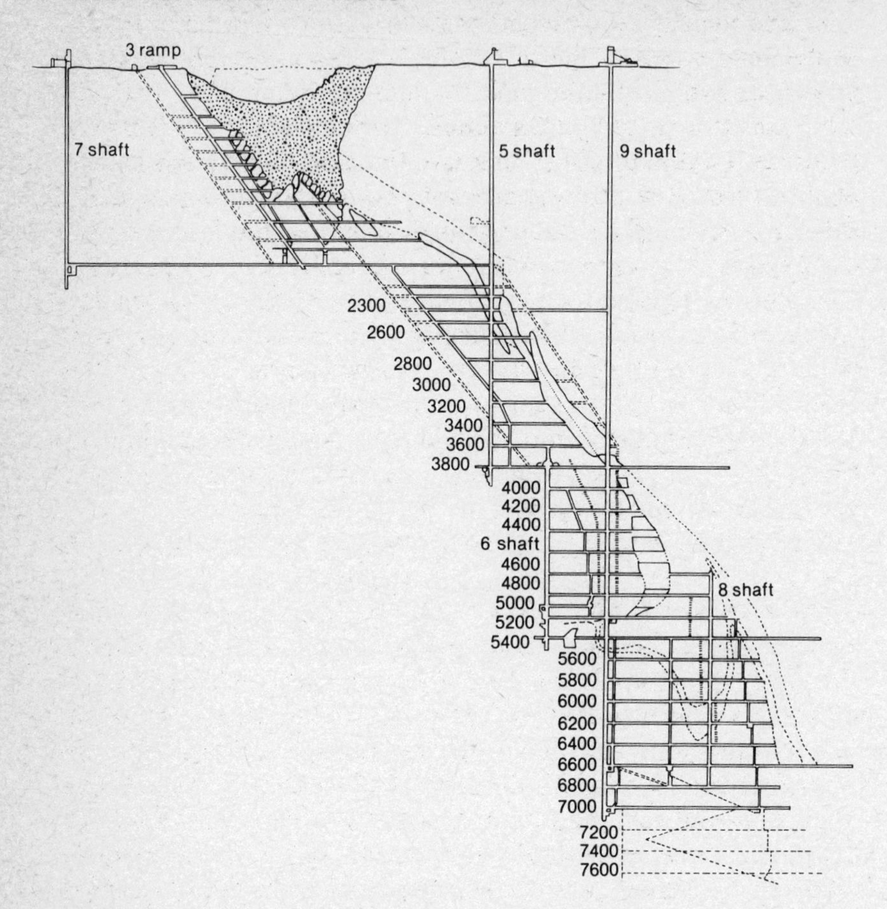

Canadian Mining Journal 98 (May 1977):24

cool it for the lower levels of No. 9. This $27-million project is scheduled to be completed in 1980, with a twenty-one-foot-wide shaft extending to the 6,000-foot level.

Reflecting the depth of the mine, the cage here is the largest in the Inco mines. It weighs 21,718 pounds and has a double deck that can hold forty-eight men or 16,500 pounds of equipment on the top, or 21,000 pounds on the bottom, with a maximum of 24,000 pounds. It takes the first ninety minutes of each shift for the cage tender to bring the men to surface and deliver a new crew to their levels. The cage tender is in communication by telephone at each level with the shift boss (who oversees the hoist) about what work is to be done. As this is a partially mechanized mine there is a lot of material to move to surface and between levels: tram locomotives, ore cars, fuel, timber, mucking machines, crusher parts, tools, spare parts, load-haul-dump machines, raise borers, metal screening, roof bolts. The cage tenders are in contact with the hoist man through a bell signal system, as in all shaft mines.

The ore in the mine is not a standard product. There are great variations in the quality and in the settings from which it must be removed. About three-fifths of the mining at Creighton No. 9 is partially mechanized using the conventional cut-and-fill method, one-fifth is blast-hole stoping, and one-fifth is undercut-and-fill pillar extraction. In cut-and-fill, which has been used here since the 1930s, the miners work from the bottom of an ore body in stopes about ten feet wide and thirty feet deep. When an area has been mined, it is filled up with sandfill until it is possible to reach the ceiling again for drilling the next layer, and the process is repeated for the entire 170-foot depth of the stope. One cut-and-fill area is on top of another. In the lower level, slushers are used to remove ore, with three slushers for two active stopes. Two slushers pull the ore into the centre column, and the other pulls it along the column to a chute. Two hundred feet below this level there are ramps and scooptrams (a kind of front-end loader; see pages 140-44). A ramp in the scooptram area provides movement between levels for the equipment, which is "captured" underground (that is, the machines have to be dismantled and brought into the area, but once inside they can move about). The men using the slushers climb about twenty feet down a passage on wooden ladders to their work sites, which are in a very old section of the mine with some buckled timbers leaving head space of less than 5½ feet. Movement between the two levels is possible by the escape shaft, which is also used to circulate the air for ventilation. The shaft itself is quite cool, but the mine is hot, particularly in dead-end areas away from the vents.

In Creighton No. 9 most raises are now driven with mechanical raise borers (see pages 137-40). The technology is borrowed from the oil industry and will eventually replace raise crews like those in Crean Hill.

Thompson Mine (T-1 and T-3)

No. 1 shaft of Thompson Mine (the first sunk) was completed to a depth of 2,106 feet in 1958 and is now 4,447 feet deep, making it western Canada's deepest production shaft. No. 3 production shaft was completed to its present 2,607 feet in 1965. As Figure 4 illustrates, the two shafts are connected underground. T-1 is used to haul the ore to surface and T-3 is used as a service shaft for men and supplies. The mine has a capacity of 7,000 tons a day, about the same as Creighton No. 9, and operates two production shifts five days a week. It normally employs about 950 men* and had a turnover rate in 1977 of 118 per cent. There are about fifty supervisors and fifty mining engineers and geologists attached to the mine.

Thompson is a cut-and-fill and pillar mine. About half of the underground employees are on bonus. They produce about 10 tons of ore per man-shift, in contrast to the 200 to 250 tons per man-shift taken from the open-pit Pipe Mine (see pages 110-12), some twenty miles southwest of Thompson. The Thompson Mine, however, has high-grade nickel ore, and the very irregular ore body lies in narrow discontinuous patches in an area one mile wide and three miles long.

Underground there are fluorescent lights in some of the main travel ways (although the stopes are not lit), which is quite unusual in mines, and concrete has been sprayed on some of the walls of these main areas to prevent the fall of

* Many men leave in the summer. In July, 1977, there were 950 men working in Thompson Mine, but by August the number had dropped to 885. The company was taking on few new hires because of impending layoffs.

rock. The mine is fairly dry, drained by ditches running along the main passages. Working conditions vary a good deal in the eighty to ninety distinct work places. Beginning in 1967–68, diesel equipment was introduced, mainly in the form of scooptrams to replace slushers in the stopes. About fifty of these are now in operation. As at Creighton No. 9, these machines are captured and are therefore not in constant operation. The midway mechanization of this mine is reflected by the fact that raise borers are used to drive many passages but jumbo drills (see page 133) are not used to any great extent.

One enters the stopes in the mechanized cut-and-fill areas (see Figure 4) by climbing about fifty feet up a set of wooden ladders (some of which are on an angle). Before scooptrams were introduced the stopes were 150 to 300 feet long and 200 feet deep, but they have now been enlarged to an area 800 to 1,000 feet long and 400 feet deep and at any given time about 20 feet high. When a section has been mined out and the ore passed down, it is sand filled and the next layer above is mined.

* * * * *

A Thompson miner's first day:
I was frightened as hell. They treated you like slave labour. I had never been subject to stuff like that before, and I had worked in paper mills, logging camps, and I never saw people treated like that. . . . They put us in the ditch, and it's foggy in places underground, especially in the fall with the difference in air. And they were sand filling – that's backfill, after they mine the place out. They put the waste back in the mine to hold the stopes up. They were sand filling, so there was water coming from the chutes, but I didn't know that. I didn't know what it was. So there were these tremendous hundreds of thousands of gallons of water pouring down into the drift where you are trying to shovel, and you are standing in water up to the middle of your shovel. I thought I was going to be drowned. And the other guys thought, "The mine is flooding; Christ, we're going to get drowned." We started to get concerned, and then they started blasting at the grizzly and the whole mine would shake. So you can see,

Fig. 4. Thompson Mine, longitudinal section

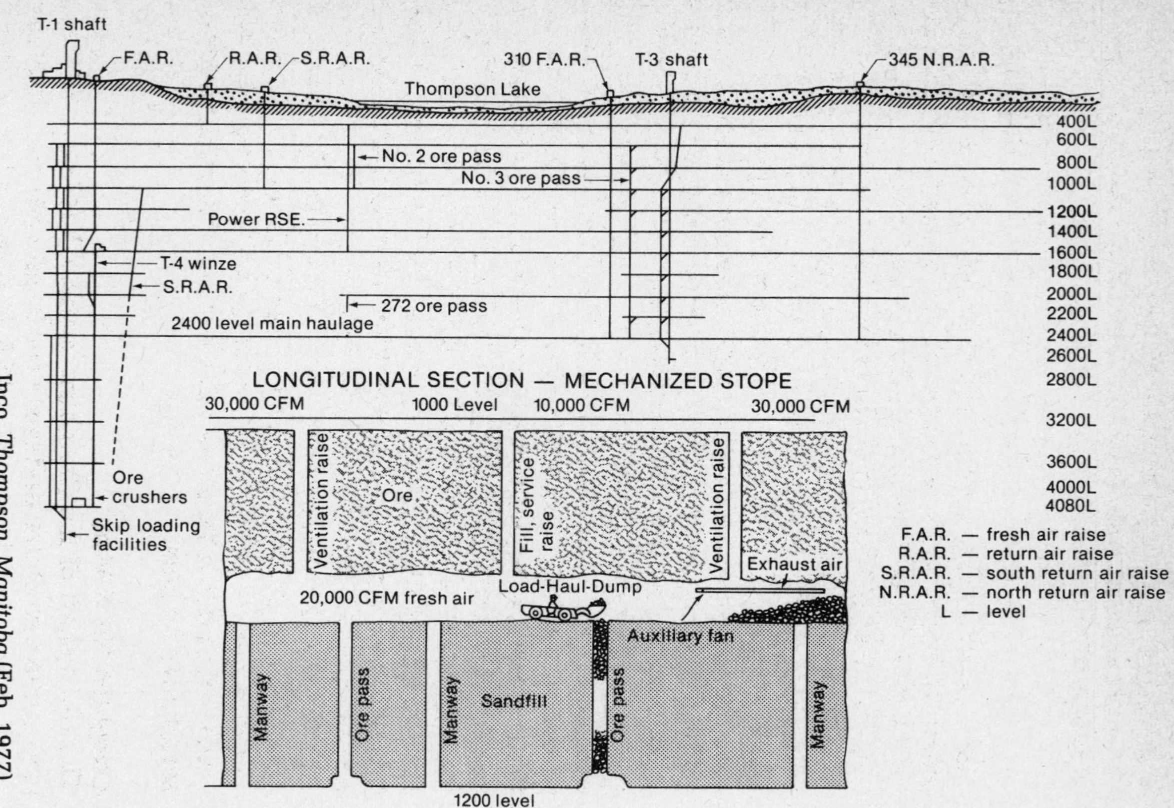

Inco, *Thompson, Manitoba* (Feb. 1977)

there was the fog, the water pouring down, and the blasting. There you were; you didn't know where you were and didn't know how to get out. And then the mine captain came along and said, "You're not working hard enough!"

* * * * *

The main jobs in a stope are those of the stope leader, a driller, a loader man, who operates the scooptram and also uses a drill, nippers, who are labourers who run for supplies, and, at various times, maintenance men who look after the scoops. On any one level a shift boss usually has production going in three stopes and visits each once or twice a shift. Usually there are about thirteen men under his supervision.

At the bottom of the ore pass from the stopes, a compressed-air chute holds the ore to be passed into electric trams (see page 145). These are operated by a two-man crew, a motorman and a switcher, who must also clear the ore pass if it becomes clogged and load the tram. They must go up a catwalk to operate the compressed-air chute and sometimes free ore that has become jammed with six-foot steel rods. The tram carries the ore to a larger ore pass and dumps it to the 4,080-foot level. Here the ore is reduced to six inches in diameter by a crusher with a 42- by 60-inch jaw operated by two men. These men also pull out, by hand, any steel rods that have become mixed with the ore and toss them into bins. From here the crushed ore is discharged to the 4,160-foot level, where it is automatically conveyed to a hoist (see page 146). At the conveyor, one man works at a control panel operating the conveyor belt and the skip to surface while another again removes scrap steel by hand.

Creighton No. 3 Mine

The No. 3 shaft of the Creighton complex has been converted into a ramp operation and now uses large-scale blast-hole mining methods, in which no backfill is required. Its output of 4,000 tons a day, about the same as Crean Hill's, is shipped to Clarabelle Mill. On each of the production shifts there are ninety to one hundred men and thirty on graveyard. The latter do service and tramming but no development work.

The mine is entered by a ramp from surface, and the men are transported by diesel trucks. The supervisors drive small jeeps, known as uni-mogs, from location to location within the mine. Compared with a mine like Crean Hill, ramp mines are spacious and allow a great deal of mobility. The ore in Creighton No. 3 is low grade, and formerly the mine was losing money; but with the ramp and use of scooptrams it can now pay its way.

At the 980-foot level six men operate 6½-inch in-the-hole drills (see page 133) in preparation for blast-hole stope mining, in which ore drops to lower levels. After 1973 a special compresser was set up here to power the drills. The area is without lights and tends to be damp.

As will be seen, blast-hole mining requires a great deal of drilling and large-scale but infrequent blasting. Although this type of bulk mining is physically much easier for the miners than the traditional forms, dangerous conditions may arise when the ground shifts because of the large "glory holes" the blasts leave, causing disruption of the mine's formation.

The material handling is highly mechanized. Scooptrams with four- and eight-yard capacity are used (two ST-4s and seven ST-8s). With a slusher, one man used to be able to move thirty to forty tons of ore a shift, but one man on a scooptram can move five hundred to six hundred tons a shift. Mechanization has made a big difference here. Unlike the practice in captured equipment mines such as Creighton No. 9 or Thompson, the scoops are in constant use here, and the driver does nothing but move ore. Another recent innovation has been the use of jumbo drills (see page 133) for development work in the mine. Creighton No. 3 has two hydraulic three-boom jumbos and one pneumatic three-boom jumbo, first introduced in 1974. The men operating these machines tend to be qualified for a number of other drilling jobs before they bid to the jumbos. Operating one of these machines is a very noisy job, and the driller works alone at the same piece of equipment for the entire shift.

Mechanization of course requires maintenance. It was possible to introduce a centralized garage here because the ramps allow the easy movement of equipment between levels. The garage (see Figure 5) has five working bays to service the forty pieces of diesel equipment that have been introduced

Fig. 5. Creighton No. 3's underground garage

Canadian Mining Journal 98 (May 1977):15

since 1968. It is well lit and roomy and looks much like a garage anywhere except that it is underground, 1,500 feet below surface. It has concrete floors and concrete sprayed on the walls. Maintenance crews work three shifts. There are forty men on the day shift, twenty on afternoons, and a skeleton crew on graveyard.

As an indication of the degree of mechanization, of the 435 men on payroll, 130 are in surface maintenance, 65 are in underground maintenance, and 240 are "operating." A major part of the work of those operating is development, preparing the way for the production workers.

There are some cool spots in the mine (45° to 50°F. in the summer); the mandatory safety glasses tend to fog up as you drive along the ramp, and the temperature changes rapidly. The cold is caused by an underground ice field between the surface and the 700-foot level, where water seeped into the old open pit and froze (see Figure 3). It remains frozen throughout the year, and cold from it causes fog patches within the mine. As mentioned, Inco engineers hope to use this ice to cool ventilation air for Creighton No. 9.

Levack West Mine

This is the newest and most modern Inco mine, located thirty miles from Sudbury. Development work began in 1970 and production in 1974. Levack Mine proper was started in 1900, uses the old technology, and currently employs about a thousand workers to get five thousand tons a day. Levack West, on the

Fig. 6. Levack West Mine, longitudinal section

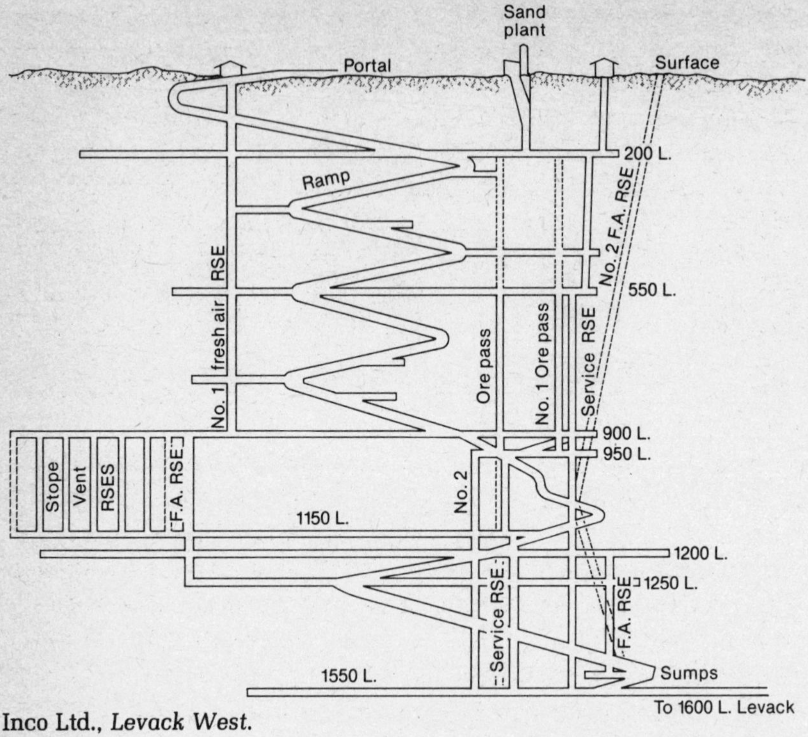

Inco Ltd., *Levack West.*

other hand, has 185 men working two shifts and a skeleton graveyard shift and produces 3,800 tons a day. Its production is about the same as that of Creighton No. 3 and Crean Hill, but it is highly mechanized, with a total of forty-seven pieces of diesel equipment in use. The result is a lot of fumes in the passageways and a good deal of dust and noise. There are about fifty men a shift working underground and of the total of 185 at the mine, 125 are operating personnel. There are also about thirty-two on maintenance. Like Creighton No. 3, Levack West is entered by a ramp (see Figure 6) which has a 20-percent switchback decline for 8,000 feet to the 1,600-foot level. The ramp itself, sixteen feet wide and fourteen feet high, has a signal-light system that is used to control the two-way traffic.

The men who work in this mine on the whole have both experience and seniority, and, contrary to custom in the other mines, it is the most senior men who operate the moving equipment. This is because the bonus system applies universally to all who work underground and has done so since the mine opened. In traditional mines older men will often bid to less physically demanding non-bonus jobs later in their careers, but here they are able to stay on bonus while doing less demanding work. They recognize the work as less skilled and less autonomous but are not compelled to take a pay cut. The average annual bonus is almost $8,000, a significant inducement for the most senior men to bid to Levack West. Because the equipment is so mobile it is difficult to measure individual output, but indications are that the universal bonus is actually designed as an experiment to end the bonus system as we now know it (see page 279).

The main method used at Levack West is mechanized undercut and fill. The cut-and-fill stopes are 40 feet wide and 100 to 250 feet long. Each blast removes a slice about ten feet thick off the end wall. Before filling, each mined-out stope is twenty feet high. There are twenty-foot pillars between each stope. Work is done in stoping blocks 250 feet wide, so that six stopes may be going at once. It takes about three months to mine a block, and for this reason all the phases of work in a complete cycle may be taking place at one time. For example, there may be drilling in one stope, bolting and screening in the next, mucking (or removal of ore) in another, and sand filling in yet another. Two two-boom jumbo drills are used to drill ten-foot-deep upholes, which are then loaded with explosives. Each 250-foot stope is blasted in three or four blasts. Next, the scalers (who knock down loose rock) and bolters stand on the muck piles and, drilling over their heads, fasten four-by-four-inch No. 9 gauge wire screen to the ceiling with six-foot roof bolts. This is to prevent loose rock from dropping and injuring those working in the stope (see page 121). The work is done by two men wh unlike most workers in the mine, have a portable lighting system. For the removal of ore, 3.5-yard Caterpillar front-end loaders are used to load huge Jarco trucks, which haul the ore to ore passes. These thirty-foot haulage trucks, although capable of transporting twenty tons, are highly manoeuvrable because they are articulated and built close to the ground. After the area is cleared, a crew of two sand fillers fill the area, putting in the floor for the next cut, and the cycle is complete. Each crew has a specialized task and does the drilling, the screening, the hauling, or the sand filling. The ore passes to the 1,600-foot level where it is moved by Grangesberg train cars (see Figure 18, page 147) to Levack Mine for hoisting to surface.

Pipe Mine

Pipe Mine is an open-pit operation located about twenty miles southwest of Thompson. Workers commute to the mine from town either by car or on a bus that passes through the city. Below the open pit is a traditional shaft mine that has been developed to extract ore below the level feasible for open-pit mines but is not yet in production. Open-pit mines are commonly used to extract ore near the surface, as was the case at the Creighton complex. Production started at Pipe Mine in 1970 after work begun in 1967 had removed 12 million tons of overburden. The giant pit's opening is 2,200 feet by 1,600 feet and its depth is now at the 590-foot level (eventually it will be 720 feet deep). The mine is shaped like a cone in the earth's surface, with benches every forty feet and a ramp that spirals at an 8-per-cent grade along the sides.

There are about thirty-five production workers on each of the two shifts and about thirty-five maintenance men who usually work days, with a skeleton crew on the off-shift. In the summer of 1977 about six women worked in the pit, a decline from a high of about twenty when the mine was on three shifts. No women now hold bid jobs (those requiring qualification and seniority) on trucks because of their relatively low seniority.

Most of the work in the mine is operation of over-sized equipment. The two 45R rotary drills and two 151 Marion shovels with eight-cubic-yard front-end buckets are all electrically powered by cables from surface (see Figure 7). Diesel equipment includes two 475 B-A Michigan bucket scoops, four D-8 crawler Caterpillar dozers, a rubber-tired dozer, and two 14E graders. Ore and rock are transported in thirteen sixty-five-ton, 700-horsepower Terex haulage trucks. All this equipment, aside from the Caterpillars, has enclosed cabs. For wall control to prevent rock slides (which is done on weekends) there is an on-going program of scaling done by a crew of five to six men using a Condor (or basket lift) and a slusher, much like those used underground.

Fig. 7. Giant electric shovel capable of loading 800 tons of material per hour

IN Manitoba (March 1978)

Drilling is done with the rotary 45R drills, which bore 8¾-inch holes in preparation for blasts of fifty-foot slices. Secondary drilling, done with mobile drill units with twelve-foot booms, reduces oversized rocks that are too large for the trucks and shovels. On the shovel there are two men, an operator and a helper who does the greasing and moves the bucket's electrical cable. A geologist marks out which piles of rock and ore are to be moved. This is usually done about six times a day. The Michigan scoops have the advantage of being flexible while the bucket is stationary but works faster because of its bottom dump. The shovels and all loading equipment have radios for contact with each other and for calling maintenance.

It is mine policy to get equipment to the shop for maintenance, but all shovels, scoops, and drills must be maintained directly in the pit. All equipment on tires is stored at the top and serviced there. (A tire contractor is responsible for the frequent repairs to the tires.) The machines are worked hard, and normally 30 to 40 per cent of the trucks are under repair. Each truck carries twenty to twenty-five loads a shift, taking fifteen to twenty minutes for a round trip.

Operators of heavy equipment are subject to a lot of jolting (especially the Cat operators), which tends to cause back ailments. The jobs are on the whole boring, and monotony is a problem, especially on graveyard shift and in those jobs requiring continuous driving up and down the pit. During the winter it is dark by 4 P.M. so that a lighting system on the ramp is necessary.

On surface, two men tend the waste dump area where rock from the mine is deposited. One works on a Caterpillar and the other guides the trucks to dump loads. The latter also watches the generator used to light the area. Trucks must stay ten feet back from the dumping edge. One truck has gone over, falling forty feet. Although the truck was smashed, the driver was not seriously injured (but the accident cost him his job). Ore is hauled to a fifty-four-inch primary gyrator crusher tended by one person. It is then fed to a vibratory feeder, conveyed to an 8,000-ton bin, and finally loaded onto Inco-operated railway cars for shipment to Thompson.

Unlike those in the other types of mines, workers in the open pit are not on bonus. Their work is always visible to the supervisors, and because it is very mechanized they have little control over the rate of production. The absence of bonus and the presence of women are the two main social differences between the open-pit and other mines. Otherwise it is similar in scale to the ramp mines such as Levack West or particularly Creighton No. 3, because of its blast-hole mining.

Innovations in Mining Methods

Over the past decade the methods of hard-rock mining have been revolutionized. In metal mining the output of the ore increased by 114 per cent between 1964 and 1973, and the value of this output increased by 158 per cent. Yet there was an increase of only 15 per cent in the labour force; each worker accounted for $54,000 of capital stock in 1964 (measured in 1961 dollars) compared with $88,000 in 1973.[1] The peak year for technological change was 1970, when there was widespread adoption of high-volume technology, originally designed for open-pit mines, in underground operations. Most important was the introduction of trackless mining machinery, especially load-haul-dump scoops (scooptrams) and mobile multiboom one-man drill jumbos (or jumbo drills), along with mechanized raise-boring units and in-the-hole drills adapted for sub-level caving (blast-hole mining). Generally these developments have meant an upgrading of equipment and production to increase the profitability of mining. The ramifications are far reaching, as the *Engineering and Mining Journal* commented: "Application of open-pit blasting techniques underground using large diameter, low profile drills will drastically alter stoping methods. Such changes in turn will affect loading, hauling, and hoisting systems, and will call for reconsideration of all mining requirements, from ground control to ventilation."[2] Mechanization has been a response to a number of factors – the grade of ore, the cost of power, the demand for increased output – and, as the *Northern Miner* observes, the "growing shortage of both skilled and unskilled workers makes it necessary to increase the degree of mechanization in order to increase productivity per man-shift."[3] But militant labour, their wage and safety demands, and labour turnover were equally important motivations behind the current developments.

To understand the impact of mechanized methods it is important to examine the mining procedures they are replacing. Some of the older methods, for example the square-set timbering described below, continue to be used. Scooptrams have been introduced into some cut-and-fill stopes simply to replace slushers, but in other cases entire mines have been designed around trackless equipment, as in the blast-hole and open-pit mines.

Square-set Timber Method

Square-set timber mining (see Figure 8) is a means of bolstering the ceilings and walls of the stope and extending the support of the ceiling as blasting enlarges the area being mined (the ore is removed from over the miners' heads while they work on a wooden platform covering the sand-filled area mined out below). This method is versatile and applicable to a variety of ore bodies regardless of their irregularity and size. It was (and in some mines still is) used when the mine walls needed to be supported to prevent caving during the extraction of ore. The following account of the procedure relies on the recollections of a recently retired Inco miner, who spent thirty-seven years in the mines, and is supplemented by a 1946 mining journal account.*

The stopes in the square-set timber method are generally five sets wide with pillars four sets wide between each stope (a set is two posts and one timber resting on their tops). The pillars are bodies of ore between the stopes left as columns to support the mine. They are mined (or recovered) when the rest of the ore body has been removed. In the stope there is a post every 5½ feet with a timber resting across the top, creating a complete network of supporting timbers. From the outside post on one side to the outside post on the other side in a five-set stope the distance is 27½ feet. The breast, or work face of the stope, is directly in front of the miner as he makes "cuts" from the ore body across the length of the stope. According to the retired miner, "If you were good, you could blast the whole five sets at once. Sometimes, if you saw you weren't going to be able to make the whole blast, you blasted three sets, or half way across, and the other shift would come in, clean up, and they would drill off and blast the other two sets. But the cycle was to drill and blast room for five more sets of timber."

As is evident in Figure 8, there were chutes in the floor of each stope. These chutes were spaced every twenty-two feet along the centre line of the stope when mucking (or shovelling of ore) was done by hand and wider apart when slushers† were introduced. A slusher is being used in Figure 8. The veteran miner said, "Mucking was all done by hand in 1940 [in Frood mine]; they had no slushers – they were just experimenting with them – but two shovellers were on the mucking floor (just below where the drilling was being done). We had two floors open, the mining floor, where they were doing the drilling and the blasting, and the floor below, where the broken muck fell down." On the mucking level the ore that had been

* The common mining methods used at Inco up to the Second World War are reviewed in detail in a special issue of the *Canadian Mining Journal* (featuring the International Nickel Company of Canada) in May, 1946, especially pp. 343–65. There were many variations in the techniques used at the time, but the most common was the square-set timber method.

† A slusher is a mechanical device for moving ore. It consists of a heavy metal blade attached to two cables, one fastened to the wall behind the pile of ore to be moved and the other to an engine that pulls the blade back and forth to move the ore along the stope to a chute.

Fig. 8. Section through a standard square-set stope

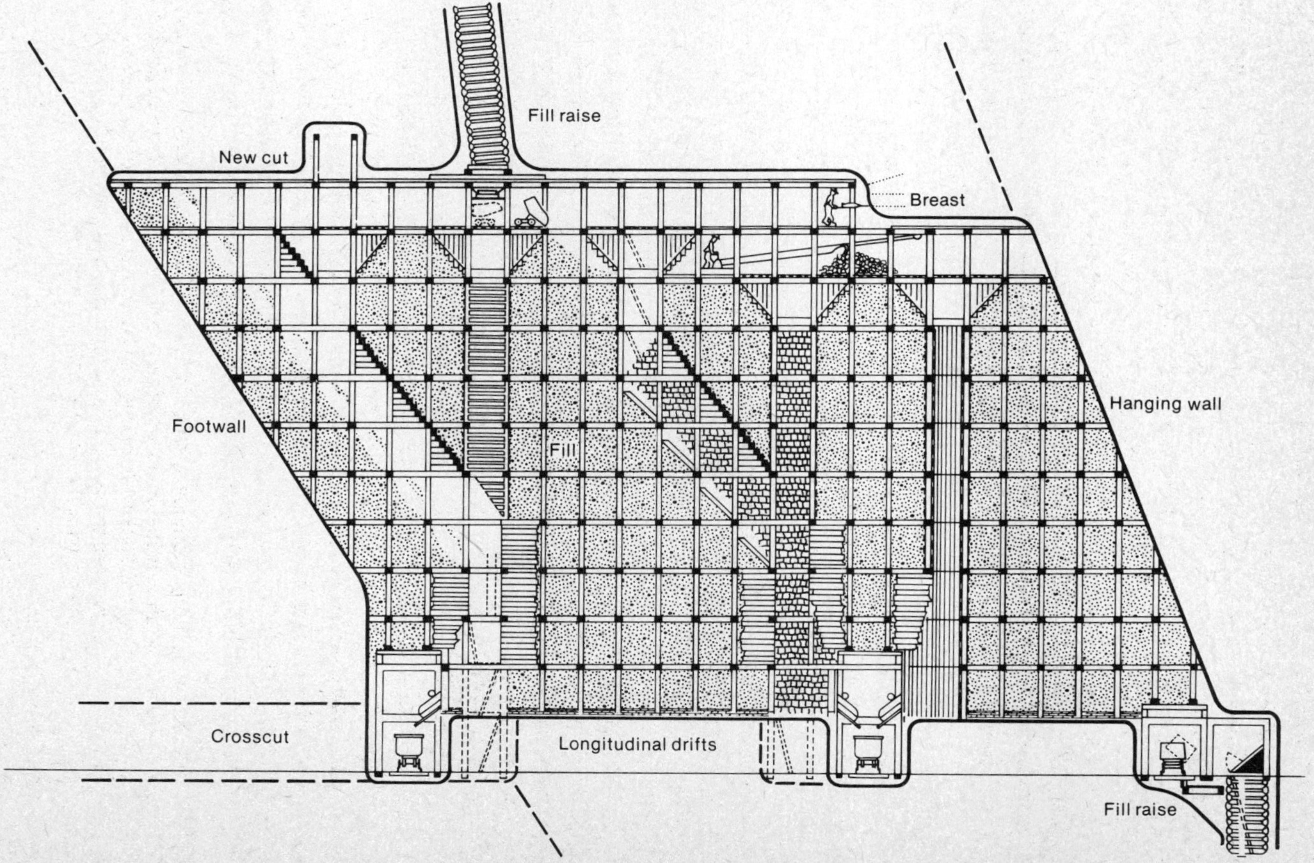

Canadian Mining Journal 67 (1946):346

blasted away from the breast was pushed down an ore pass. This chute had an opening about 4½ feet square covered by "grizzly" rails joined with four or five short pieces of railroad rails, leaving a number of thirteen-inch openings. The ore had to be broken fine enough to go through these openings.

* * * * *

At that time [about 1940] *most of the work was straight hand work; the only machinery was the drills, the drill for drilling in the breast and then a small one, what we called a plugger, the same thing as these pavement breakers out in the street, for drilling chunks that were too big to go down the grizzly rails into the chute. . . . We had to break up the muck fine enough to go through those rails in a thirteen-inch space. However, it took a lot of work to do that, and any crew that took the time to do that were just as liable as not to be told, "Look, you are no good around here. You can go to surface and get your time and go find another job." So, what we used to do, just after lunch or just before quitting time, was knock a grizzly block apart and spread the rails wider so we could roll the big chunks in. I was doing that one time, and I got a chunk stuck between the spread rails. I was up two hundred feet above the bottom of the chute. As I was working to shove this thing down, I heard the voice of this safety engineer up on the mining floor, talking to the stope boss and the driller. I worked at that chunk and finally got it down, got the rail back in place and the grizzly back, and as I turned for another shovel of muck to put down the chute the safety engineer came down. I had only been there about a month or so, and still didn't know that there was a safety belt I was supposed to wear when the grizzly was broken.*

* * * * *

The mining cycle for square-set timbering was quite complex. There were usually two production crews working the day and afternoon shifts and a service crew that brought in timber and drills working the night (or graveyard) shift. A production crew usually included a stope leader who was respon-

Fig. 9. Scaling

IN Manitoba (July 1972)

sible for organizing work, a driller, and several shovellers (later replaced by a slusher operator). When a crew entered their stope the previous shift had probably just completed a blast. While some members of the crew set to work getting the ore (or muck) down the chute, the rest would begin to put up timbers, clean up the mining floor, lay down wooden planks on the floor, and scale* any loose rock in the area just blasted (see Figure 9). After "booming out" heavy timbers over this area to prevent any loose from falling, they would begin to move the muck pile away so that a base could be made for the drill. Depending upon the amount of drilling required, the driller or the driller and stope leader would begin to drill while the others moved the muck down the chute, finished blocking and bracing the timbers, and prepared for the next blasting. Finally, someone was sent for blasting powder while others removed the drill and checked the holes before loading them.

* Scaling is the knocking down of loose rock that may have been freed during a blast but has not fallen. The scaler first knocks on the walls and ceiling of the work area with metal bars from six to fourteen feet long, listening to the "ring" of the rock. "Loose" is then wedged down using the chisel head of the scaling bar. Loose is any rock that may fall, regardless of size, and may weigh a few ounces or a few tons.

* * * * *

We used dynamite at that time and staggered fuses in order to get it to explode in the proper rotation. See, in the stopes we blast anywhere up to forty holes. I blasted 105 holes one time, used 856 sticks of powder. But as a rule, in the stopes it would be thirty-five, forty, forty-five holes and in the pillars we'd blast eighteen, twenty, sometimes twenty-five at the most.

We had some short pieces of fuse two feet long and would take a jack-knife and notch the fuse every inch or so. We would notch two of these short spitters for each of us, so we had a spare in case one went out. We would cut the fuse into different lengths so that they were from seven to ten feet long after we had the holes loaded. Then we'd light all of our spitters, and two of us would go and light all the fuses in rotation, and we'd leave the stope. By the time we had them all lit there would still be about five feet of fuse to burn. That would take about three minutes and twenty seconds. That would give us lots of time to get out of the stope into a guarding position and wait for the blast.

Later on, one of the technological changes somebody came up with was a kind of slow-burning wire, and with thirty feet of this wire you could have the fuses all the same length. The wire would burn from one fuse to the other and light all the fuses. All you had to do was light one place on the wire.

* * * * *

Square-set mining was the most extensively used method in Inco's main ore bodies; today it continues in limited use in some corners of older Inco mines, but for the most part it has been replaced. It was heavy physical work that demanded great skill and precision on the part of miners, not to mention a tremendous amount of timbering. Most were glad to see it come to an end. They preferred the newer cut-and-fill methods that continued to demand their skills but required less "bull-work" and increased their productivity.

Cut-and-Fill Method

The routine in cut-and-fill is similar to square-set mining in that there is a cycle of work made up of drilling, blasting, scaling, and removing the ore. In undercut-and-fill (or underhand mining), the ore is removed from the floor of the stope, but in conventional cut-and-fill (or overhand mining) it is removed from the ceiling. In both cases, the emptied area is refilled with sand mixed with cement to provide ground support so that more ore can be extracted. In undercut-and-fill the miners work beneath the sandfill and in conventional cut-and-fill they work on top of it. Open stoping, which does not use fill (as in the blast-hole method to be examined shortly), is used when the surrounding rock is strong enough to support large cavities. It is obviously more economical than cut-and-fill but cannot be used for many ore bodies, particularly such irregular ones as those in Thompson and many in the Sudbury area. The cut-and-fill method is used in most of Inco's mines, including traditional hand mines, captured equipment mines, and ramp mines. The actual work area can range in size from the rather small hand and captured-equipment stopes of Creighton No. 9 to the large captured-equipment stopes of Thompson Mine right through to the giant mechanical stopes of Levack West. The technique that they have in common is that ore is removed in horizontal slices and the empty area is filled before the next slice is mined, with the fill supporting the stope's walls.

Cut-and-fill of the conventional type was introduced by Inco at the original Creighton Mine in Copper Cliff in the 1930s. Not until the 1950s was the undercut-and-fill method first used by Inco, this time in the Frood Mine near Sudbury. Stopes in the cut-and-fill method are arranged in a sequence along the ore body with pillars left between the stopes for support. Later, also by cut-and-fill methods, these pillars are mined (or "recovered").

The following account of the introduction of undercut-and-fill mining is given by a Sudbury miner who took part in Inco's first experiment in the mid 1950s. Figure 10 graphically depicts the technique he describes.

The foreman I was working with, he spoke up. He says, "Listen, I have a pillar that is just about ready to mine. Suppose, instead of mining it overhand, we try it underhand." So we started out. We mined the first cut, and the second cut, and the third cut, and by the time we

Fig. 10. Undercut-and-fill method of mining

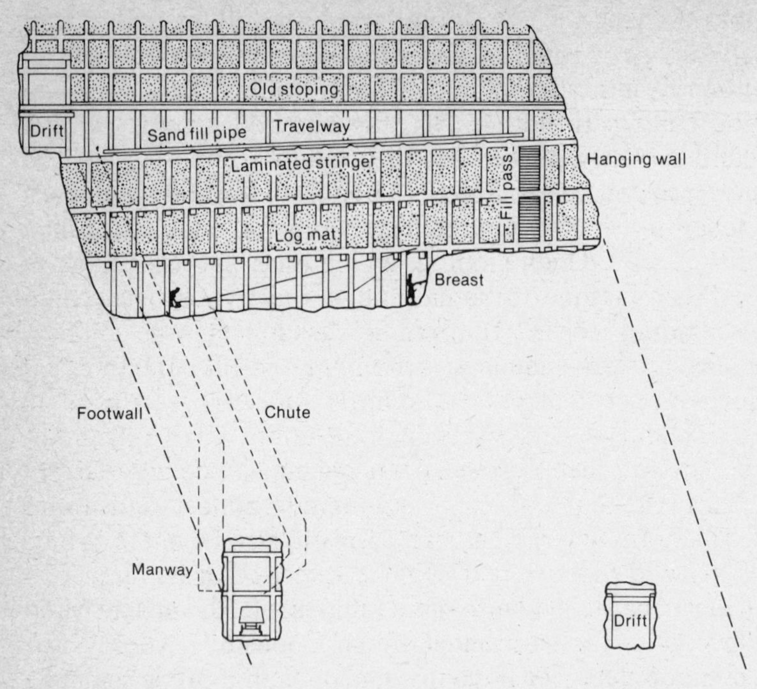

F. B. Howard-White, *Nickel: An historical review.* Toronto: Longmans Canada, 1963.

got done mining the third cut, we were going great guns. We only had seven floors in that pillar, and by the time we finished it, word went around that all timbered areas were going to be gradually changed over into this new undercut-and-fill method. Now that was the first advance in timbered mines, the first real big change in one hundred years, and it was started right out in Frood Mine and I had the pleasure of being in on it.

All the old areas that are broken up, they changed them over into this undercut-fill method. It increased the output and it certainly cut the cost for timber. Before, there had to be a post every 5½ feet both ways, and heavy timber across both ways. Well, now they cut out fan-shaped posts completely. All they do is post it up with round posts, just sawed in the bush, cut off top and bottom, limbed, and sent down underground. Very cheap. There is no sawmill work to frame these posts. The stringer [placed on top of the posts] *is made of two pieces of five-inch timber bolted together. This stringer would be pushed forward and then another section would be clamped on – a simple stringer and then this plain log post underneath. The savings in timber must have been enormous, and the Frood Mine timber yard, three-quarters of that is vacant now, and what there is is mostly this simple timber.*

Sandfill is carried in the mines by four-inch pipes that can pour two hundred to three hundred tons an hour. Before it is poured the miners must build extensions to the ore passes and manways. At one time a sandfill crew known as a pipe gang prepared the sandfill pipes, which were then made of heavy cast iron and required five or six nuts and bolts to fasten them together. Now thin-gauge steel pipe is used and can be fastened with a clamp and two bolts so that sandfill crews have less responsibility for setting up the pipes.

Another innovation, the rock bolt, made possible the widespread use of cut-and-fill methods. It takes the place of timbering in preventing rock from falling into the stope as loose. Experiments begun after the Second World War found that a bolt can support a great deal of weight: an eight-foot rock bolt can hold eight tons of ore or rock. They actually hold better than the traditional timbers and thus improve the structural support necessary for cut-and-fill mining. A bolt is made by drilling a hole into the ore or rock and inserting a long (six- to twenty-one-foot) steel rod into the hole with an expansion shell on the end to anchor it. On its head is placed a large washer, and pressure is applied to firm up the bolt. Rock bolts are often used in conjunction with a heavy-gauge wire screen which retains smaller pieces of loose.

As mentioned earlier regarding Thompson Mine (see page 103 and Figure 4), there were important changes in the scale of cut-and-fill stopes after captured scooptrams were introduced. The stope size increased fourfold and the crew size grew from two to four or five men. Yet most of the men still work in the various phases of the mining cycle and tend to have a great deal of control over the pace and direction of their own work. Cut-and-fill mining requires a great deal of skill learned gradually by working closely with partners in "your" stope. The miner has to know how to drill and blast in such a way that the ore is broken up without blowing everything else apart; he has to be able to operate a slusher or, in a mechanized stope, a scooptram; and he has to know how to set up stringers and prepare for sandfill. All this takes experience and knowledge, not just hard work. It usually takes a miner about six months of using this method before he earns a bonus, no matter how hard he works. On the other hand, as we shall see shortly, in the very large cut-and-fill stopes at mines such as Levack West, each of these tasks becomes so specialized and sub-divided that some men do only drilling, others only bolting, and still others only operation of scooptrams. Technically the method is the same, but the organization of work changes radically when the stopes are enlarged and access is made easier by ramps.

Blast-Hole Method

Since the early 1950s, an "induced caving" method has been used by Inco, particularly in Copper Cliff North Mine, the Stobie and Creighton mines in Sudbury, and more recently between the 1,900- and 2,300-foot levels of Birchtree Mine in Thompson. It was the first use of open-pit mining or surface techniques underground. A second generation of this method is blast-hole mining, which differs from its predecessor only in that more explosives, loaded into larger holes, are used to break the ore during blasting. Blast-hole mining is actually simpler from an engineering standpoint than other methods; according to the *Engineering and Mining Journal*, "blast-hole layouts are much simpler, reducing engineering time about one-fifth of that for longhole drilling; better fragmentation is achieved, with a reduction in secondary blasting and an improvement in drawpoint efficiency."[4] As is evident from Figures 11 and 12, blast-hole mining involves opening drifts at the top and bottom of an ore body. The drilling takes place at the top and at the bottom the ore is removed. Vertical blast holes are drilled, the holes are loaded with explosives, and the blast throws the ore into the open slots.

Fig. 11. Blast-hole layout

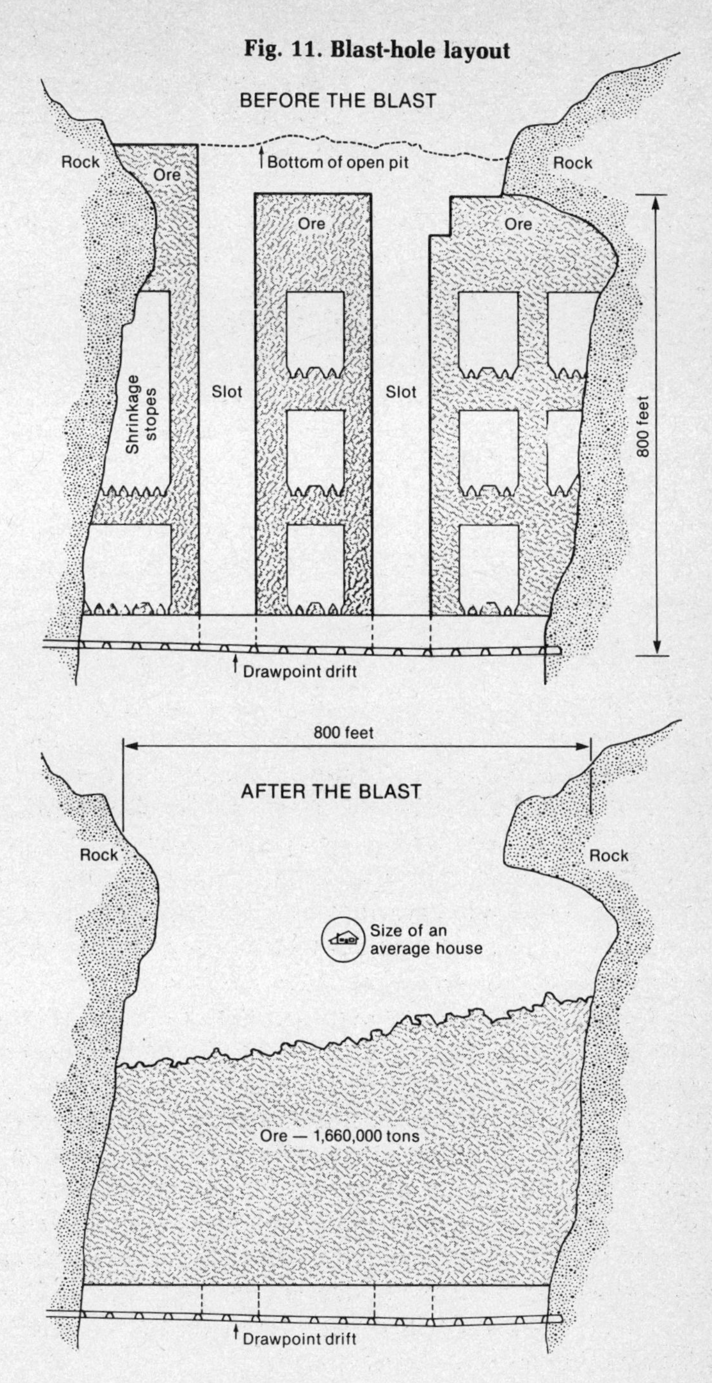

Inco Triangle (May 1974)

Fig. 12. Blast-hole bench blasting

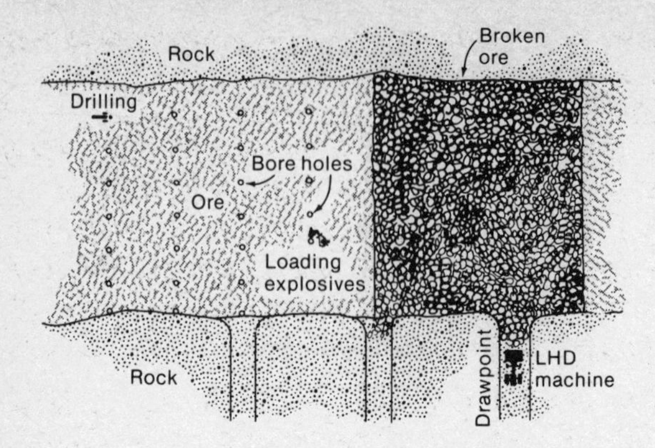

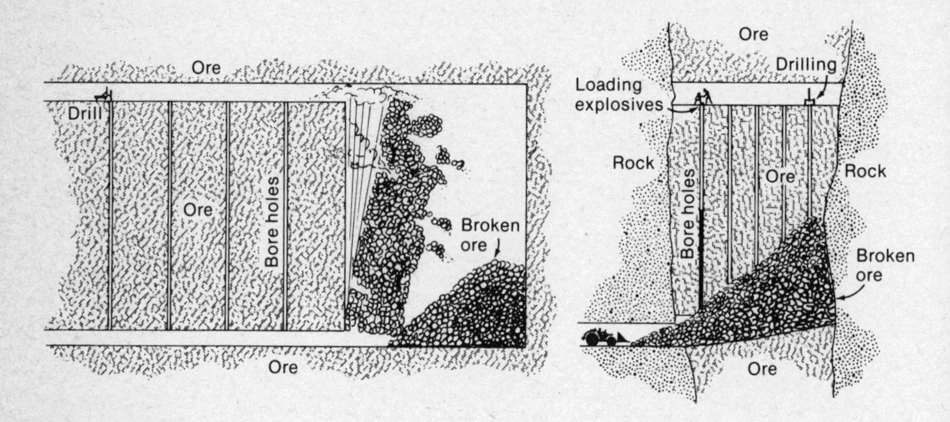

Inco Triangle (October 1973)

* * * * *

Observations of a Sudbury miner in a mechanized area and another in a traditional area:

Miner 1. *In a "go-go" area you can end up doing the same job week after week after week, like putting in bolts and screens. In conventional mining you would drill for a while, blast, timber, roof bolt, and there were a variety of tasks; but now one guy does his task. His task might be roof bolting and another guy might be running a jumbo. A couple of other guys might be running the scoops.*

Miner 2. *The undercut-and-fill place where I work has just two men, the two guys working together, even the driller. You mine for three weeks or more, and then you prepare for sand.*

Miner 1. *The guy in the scoop area, he knows that that's what he's always doing. He's going to bolt, to bolt, to bolt. Another guy knows that his job is going to be to load, to load, to load, to load. You never stop; it's just one job after another.*

* * * * *

The blast pictured in Figure 11 took place at Copper Cliff North Mine in March, 1974. Over 260 tons of explosives were loaded into 12,936 drill holes (116 miles of them), and the blast produced 1,660,000 tons of ore. This was a particularly large blast. In contrast, at Birchtree Mine in Thompson an average blast or horizontal "slice" produces about 5,000 tons of ore and uses about two tons of explosives. The scale of the blast-hole method, compared with other underground mining techniques, is nevertheless enormous. It is more economical than the others in terms of labour power required. It is only practical, however, where ore bodies are large and regular, and it demands much development work. The scale now used by Inco also requires a great deal of mechanization, particularly at the drilling and draw points. The machines used in these operations are the in-the-hole drill (shown in Figure 15) and the load-haul-dump or scooptram (shown in Figure 17). Unlike the captured scooptrams in the cut-and-fill stopes, these are in continuous operation in blast-hole mining, and their drivers do nothing but move ore from the draw point to an ore pass. In the earlier induced caving mines the ore was moved by slushers and trams, as an old Inco miner recollects:

> *When these long holes were first started, they would drill a drift across the ore body and drill rings of holes from one end of the drift to the other. They had a slusher trench below these and box holes from up above running into the side of the slusher trench, and they brought in great big slushers, 125 h.p., with big scrapers weighing about ten tons or more, and they would slush it out with these big scrapers on the 125 h.p. electric slushers.*
>
> *In some cases you had the train just sitting on the track below you. You would slush it into the train: you slushed the muck forward near the draw hole until you got a trainload waiting there, and when the train came, all you had to do was just scrape it down into the hole.*
>
> *Anyway, this long-hole drilling made a big difference in the mine. However, in the areas that had been opened up for square-set mining, you couldn't drill with these long holes in there because the ore was all broken up so even now they are still mining in these old areas to finish them. But in the new areas of the mine they start up at surface or at the top of the ore body and they have trackless mining.*

Blast-hole mining in Creighton No. 3, the ramp mine discussed earlier, uses the most advanced techniques in this form. Drilling is done by an in-the-hole drill. About 100,000 feet of drilling is needed to produce one million tons of ore, and about a pound of explosives is required for each ton of ore. On average five months of drilling is done in preparation for a blast (see Figure 12). The holes are loaded for blasting with an orange gel-like substance known as ANFO (ammonium nitrate–fuel oil). A team of three men loads this explosive along with detonators, one man tethered by a line because he works close to the edge of the hole formed by the previous blast. None of these men is considered very skilled. The material itself is handled roughly, being tossed into the six-inch by three-hundred-foot holes. The actual loading is done by helpers, but a blast foreman is in charge. He is one of the mine foremen who oversees all the in-the-hole blasting in the mine. The large blasts occur on Saturdays when the mine has been cleared.

The equipment and explosives used in blast-hole mining were originally developed for use in open-pit blasting; the adoption of this method in underground mines is revolutionizing production. The caving technique was relatively easy to mechanize because it offered a steady drilling pattern with few variations, unlike cut-and-fill or timbering, where to be efficient the drilling techniques require highly skilled miners. In all the other methods, including the mechanized ramp cut-and-fill mines, there is the sense of a cycle of work. Either the miners themselves or others they can see perform all the tasks necessary to mine and remove the ore. In blast-hole mining they are detached from the overall project. It takes months to prepare for a blast. The miner is a driller working at the top of the ore body, a scooptram operator working at the bottom, or a worker loading explosives, without any sense of a mining crew or partners. Each phase of the mining cycle is separated from the others. There is little need for direct co-operation. The miners have very little control over the rate of production; most become machine tenders or machine operators. The logic and strategy for removing ore is no longer in their hands. It has been transferred completely to the engineers and, through them, to management.

Underground Mechanization

The three mining methods just outlined require progressively more mechanization, less labour, and fewer skilled miners. Very little mechanical equipment is used in traditional square-set timber mines, but most people in the blast-hole mines operate or tend a piece of large-scale equipment. The cut-and-fill mines range between these extremes. Some are still primarily hand labour operations, using jack-leg drills and slushers; others have captured scooptrams instead of slushers; still others, as at Levack West, use jumbo drills and scooptrams. The more mechanized equipment used, the more likely a miner is to perform only one aspect of the mining cycle, the faster he can be trained to perform his task, and the greater is the scale of the operation. From management's perspective, as more equipment is introduced they have to rely less on the skill or individual initiative of the miner. Since 1965 Inco management has introduced over five hundred pieces of diesel trackless equipment into the mines through its program of capitalization.

A few key changes in the equipment of mining can be identified, and they have to do with the way some of the important aspects of mining are performed: drilling, driving of raises, moving the ore within stopes, and moving the ore within the mine (tramming). The innovations in each of these areas will be examined in turn, followed by a discussion of the increasing demand for maintenance associated with this recently developed equipment.

Drilling

The most basic skill in mining is drilling. All hourly paid underground workers are trained on the drill before anything else. The early power drills were pneumatic, propelled by compressed air taken into the mine from surface through a series of pipes and then into the stopes, where a one-inch hose brought it to the machine. Drilling, the first aspect of mining to be mechanized, evolved very little until recently. The early "liner" drills were stationary; then more portable jack-leg drills were introduced (see Figure 13). This change did not radically restructure the nature of mining. Not until the introduction of fan drills and giant jumbo drills, followed by the in-the-hole drills for blast-hole mining, did important changes take place in the nature of the work. With these larger machines, drilling became the sole activity of many miners. Instead of being craftsmen performing a range of tasks over which they had a great deal of control, they became machine tenders.

* * * * *

A Sudbury miner who started mining in 1941 describes the early drills and some basic changes:

When I started, the drills were called liners. They were pneumatic drills, but they were set on a slide. There was a bar five feet high that stood on the timber below and

Fig. 13. Jack-leg drill

IN Manitoba (October 1972)

against the timber above. You could tighten it real tight. Then there was a cross-bar you bolted to that by means of a universal joint. Then there was what we call the dump that you put on the cross-bar so you could point your machine right up in the air or you could point it flat or point it down. And that was what we drilled with.

There was a worm on the frame, and it had a handle to turn and run this machine forward. Well, then after I had been there about three or four years, they began to put a little motor on the rear end of the machine. They called this an automatic machine because you just turned a little lever and the little motor would shove the machine forward or pull it back.

So then we gradually switched over entirely to automatic liners. The drill steel at that time was just straight steel. At the end of the shift, your steel would be dull. The graveyard shift, which was the supply shift (they didn't do any mining – the mining was all done on day shift and afternoon shift), besides nipping timber in for you nipped a bucket of steel, fresh scrap steel, and your rolled dull steel, they would nip that down and load it on the truck to send to surface. The next day that would go to* [the shop crew on] *surface, and they would heat up the ends of it and forge the bit sharp again. It wasn't really a bit, it was just the end of the steel; they would forge it in the shape of a bit and temper it again.*

Then around 1944 somebody came up with the idea of having bits that you could just shove on the end of the steel. They were a special shape, and the steel was shaped to fit inside the rear end of the bit. The bit was the same stuff, just case-hardened steel, and you just shoved them on the steel. When they were dull, you simply knocked them off and replaced them. It wasn't any use trying to salvage these bits. At that time all the steel bits were carbon steel. The engineers and scientists somewhere developed a carbide that was strong enough and tough enough that it could be used for drill bits. So they began making these carbide drill bits, and they were

* To nip is to go for supplies – blasting powder, drill bits, timber. A nipper is also sometimes referred to as a "go-for."

introduced. They made an enormous difference in mining. Up until that time, unless the ore was very soft you could only drill about six feet, maybe ten feet in some other types of stopes, and that was as far as you could drill at a time. But with the development of the carbide bits, you were able to fasten one rod on to another and keep on drilling, and you could drill four or five hundred feet with the one bit. This made it possible to drill long holes and blast them.

When you drill underground in a stope, or especially driving the drift, which is what you call a tunnel, one machine with the echoes of that noise coming from the back and the sides and the breast will sound like about four or five of these pavement breakers working on the street.

* * * * *

The most common drill today in production stopes is the jack-leg drill (as in Figure 13), used in both traditional hand mining and mechanized stopes with captured equipment. The jack-leg is so named because it has a heavy metal support at the bottom that helps steady it as the miner is starting (or collaring) his hole and provides leverage as the drill is being pressed forward.

* * * * *

A Thompson miner describes a change in the jack-legs in the sixties:

Those were Ingersoll-Rand drills and they weighed about seventy pounds, but they didn't drill as fast [as the Copetil drills]*; they were a hell of a lot lighter and were a lot easier on the man in terms of holding it up because when you hold the drill you have to start on the base, and that's a lot of weight to hold, especially in awkward places. Then they replaced them with the Copetil drills that weighed about 120 pounds, but they drill like crazy. I'm telling you, they are man-killers. Oh, they're a terrible weight, and you have to hold them up. You have to remember you are not only holding the drill but the steel too, and you have to collar in the place you start to drill. Once you get your* *hole started it's not so bad, because you can apply pressure on the leg to hold the steel and keep it from falling, but the starting is something, especially since the muck piles are slippery, and you stumble and fall all over. You are usually drilling on a cracked-up breast, and your steel gets stuck and breaks and jams and so forth. To drill in a pillar you need a fair amount of experience in order to drill through all the cracks and keep the hole open.*

* * * * *

Traditionally, in most production stopes miners not only drill but also perform other tasks required in the mining cycle. With the introduction of larger mechanical drills they operate a drill throughout the shift. The first of these machines was the mobile fan drill. It has two drills and is self-propelled. Once it is situated, hydraulic lifts are used to raise the machine, and the two "booms" or drills, which are also hydraulically operated, are placed against the mine wall. The fan driller works alone in a drift. A Sudbury fan driller said, "It's noisy. Usually you are in the dead-end drifts because you are drilling these to be blasted. Two machines in these drifts, two big machines, twice the size of the jack-leg, now that's really noisy. And there's always oil in the air with those machines. You are always wet. The water's coming down on you constantly." With the fan drills the miner drills a fan-shaped ring of holes overhead about every five feet for the length of the drift, perhaps twenty-five to forty feet. When this pattern is complete he will go into the next drift and begin again. Jumbo drillers perform the same type of work cycle, but they work in production stopes in the larger mechanized areas.

* * * * *

A Sudbury miner describes the jumbo drill:
Somebody came up with the idea of a jumbo drill, and this went back to the older type, only a heavier machine set up on booms. It had three booms on a big tractor and three machines on this. And it was made with a long carriage so that you could use ten- or twelve-foot steel in these machines. They would drive this jumbo into a drift where they wanted to drill, hook a four-inch hose on it for air,

and then the one operator would stand there with a bunch of little levers. He would operate all these three machines at once. With these new jumbos and three machines set up, all you do is simply drive the steel in, pull the steel out. A man sitting there with these little levers moves the machine down sideways and drills another hole, pulls the steel out. Some of those drillers will drill 200 ten-ton rounds a shift. They won't blast them. A blaster will come along afterwards, maybe the next day. Some of these drillers will drill off two drift rounds twelve by sixteen, which will contain about 200 tons of ore, in one shift.

* * * * *

The standard jumbo drill (see Figure 14) has two or three booms that are air-operated (pneumatic) and extremely noisy. In 1974 Inco introduced a three-boom electric-hydraulic jumbo on an experimental basis, and since then several more have been put into operation. This machine is somewhat quieter and can drill at almost twice the rate of the older jumbos. It is propelled by a four-wheel-drive diesel engine but trails a 250-foot cable that must be connected to a 550-volt power supply to operate the drills. Thus although the machine is mobile, it requires access to power sources, and these services are not available everywhere in the mines.

The in-the-hole drill is used to make the deep 6½-inch diameter holes required for blast-hole mining, as in Creighton No. 3 (see Figure 15). The operator's task is to keep the in-the-hole drill clean and keep filings (or finely ground ore) away from the drill. He adds five-foot lengths of pipe from a crib set near the drill to the drill shaft; each new length takes about one minute. The holes average 200 feet in depth or forty lengths. While the drill is working down the five-foot length, the driller simply clears filings that build up at the base of the drill and watches the machine. To become an in-the-hole driller a miner, after qualifying as a regular driller (two or three weeks), must bid for in-the-hole training. He can learn the operations in a day or two on the job.

The in-the-hole drill has many advantages: "Work areas are cleaner, there is less dust and noise, and considerable reduction in physical effort."[5] According to the *Canadian Mining*

Fig. 14. A three-boom jumbo drilling up-holes

Canadian Mining Journal 98 (May 1977)

Fig.15. In-the-hole drill

Canadian Mining Journal 98 (May 1977)

Journal, the noise levels decrease from 117 dB(A)* to 100 db(A) between conventional and in-the-hole drills. Although this is still very noisy, it is an improvement. The first in-the-hole drills were used by Inco in 1973 and have been adopted for long blast-hole mining. In early blast-hole technique, the holes were made with diamond drills. Then heavy percussion drills were used, but with these it was difficult to keep the holes straight and prevent the bits from breaking. Since 1973 Inco has introduced twenty-four in-the-hole machines, representing a substantial saving. "Direct drilling costs with the in-the-hole operations are calculated by Inco at $0.24 per ton, in comparison with $0.55 per ton for the conventional method."[6]

Figure 15 shows an in-the-hole drill. The machine is quite compact, telescoping to 5½ feet so that it can be moved to dif-

* dB(A) is the symbol for decibel. See Table 22.

ferent levels by cage. It is 4½ feet wide and 11½ feet long; the mast rises to 11 feet when in operation.

Driving Raises

Most mines are a network of drifts or levels connected by various vertical passes. The connections between levels are known as raises. As we saw in the undercut-and-fill mining method outlined earlier, the ore is moved to the lower levels by raises called ore passes, which are vertical chutes in the rock to pass broken ore to different levels for hauling or crushing. Raises are also necessary for service passes (compressed air, power, and water), manways, and ventilation. The driving of a raise is one of the most skilled and difficult jobs in mining. The number of persons engaged in driving a particular raise varies with its size: one man for a 7-by-11-foot, two for an 8-by-9-foot, and three for a 9½-by-16-foot raise. The bulkhead of the raise, or the timber at the face of the raise as it is being driven, is reached by a manway in which there are a ladder, a two-inch air line for the drill, a one-inch water line, and a one-inch air line for ventilation. Over the manway there is a bulkhead to prevent the rock from falling into it. The process is to drill about thirty-two holes, nine feet long each, into the face of the raise (over the miners' heads), load the holes with explosives, and detonate the explosives to bring down the rock. The broken rock is passed down in a passage alongside the manway. Cribbing is built against the sides as the raise goes up (to within three feet of the face). A raise cycle thus involves scaling loose rock from the face after a blast, raising the timber cribbing, clearing the rock from the bulkhead, drilling the rounds, loading the explosives, and blasting.

This traditional method is slow and dangerous work, needing up to three months to drive a two-hundred-foot raise between levels. It is still in use at Crean Hill and many other mines. One crew that we saw driving a four-hundred-foot raise for an ore pass is considered to be part of a cracker-jack team and earns 140 per cent bonus. Unlike most miners who take their break as soon as they enter the mine, this team goes directly to work and goes non-stop until near the end of shift, when they take their lunch break. When we were there they were 250 feet up into the raise. At the beginning of shift, after reaching their level they climb up the narrow manway on crooked wooden ladders to the current face of the raise. There is a good deal of dust and heat at the head of the raise, and simply getting to the work site is a difficult (as well as frightening) task in a hot, dusty, confined space. Alongside the manway is a chute. The men construct the cribbing for this raise as they work their way up, building a platform every twenty feet. Two men work at the head of the raise and another at the bottom passes up supplies. After a blast they scale the head with six-foot metal rods and shovel the muck down the chute. This work is all done at the top of the raise, the men standing on planks laid across heavy wooden headers. They then drill, set up the bulkhead, and blast (at which time they go to the bottom), afterwards continuing to build the frame to reach the top. The only light they have is from their helmets. The work is very demanding physically, and the skill level is very high. These men are greatly respected by both fellow miners and supervisors.

Recently a new method of driving raises by means of a raise-bore drill has been devised. Inco used the first experimental machine in 1964 to drill raises three feet in diameter. In 1968 five- and six-foot raises were drilled. At that time there were only about ten raise borers in the world,[7] but there were a hundred by 1972 and nearly two hundred in 1975. Even as recently as 1975 only twenty-four of these machines were in use in North America. Since their introduction some thirty-seven *miles* of raises have been bored in Sudbury. In 1977 Inco had fourteen machines operating, including some that could drill raises eight feet in diameter. An 1,100-foot drain hole was recently drilled by a raise borer in Thomson's Pipe No. 2 shaft. An eight-foot raise borer bit is shown in Figure 16.

* * * * *

An Inco shift boss in Sudbury:
The raise borer revolutionized raise driving because it's damned hard to get good qualified raise miners these days. People that want that type of work and have the ability to do it, they're few and far between.

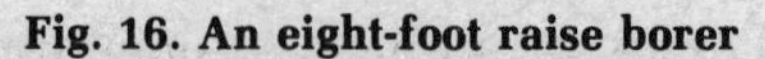
Fig. 16. An eight-foot raise borer

Canadian Mining Journal 98 (May 1977)

Two Thompson miners:
Miner 1. *I think the biggest advance made is the raise borer.*

Miner 2. *The raise borer is hard to beat. It saves a lot of danger and, whatever else, guys getting injured.*

Miner 1. *You see, drilling and driving raises are the most skilful jobs underground.*

* * * * *

Raise borers are located in "stations" eighteen feet high by ten feet square and fastened with steel posts to a concrete pad. They first drill an eleven-inch pilot hole down to the next level; once the drill breaks through, the pilot drill bit is removed and replaced by a reaming head six to eight feet in diameter, which is drawn back toward the drill machine (which looks similar to an in-the-hole drill as pictured in Figure 15, although somewhat larger). The drill pipe, which is in five-foot steel sections with a ten-inch diameter weighing seven hundred pounds each, is removed by a small hoist, and the reamer is pulled back to within seven feet of the station. The remaining rock in the section is then blasted away. Shafts are installed by machine and are lifted into place by a half-ton hoist fastened to the roof and side. The development of these machines has increased Inco's capacity to drive raises and comes at a time when fewer raise crews with the necessary skills are available (not having been trained by Inco, of course, in recent years). Operated by two men and used on all three shifts, the borer is able to drill about thirty feet a shift. Inco's maximum is now a diameter of eight feet, but the technology has been developed for sixteen- to twenty-foot raises. Over two hundred holes have been drilled using this method at Creighton No. 9. The work areas are well-lit cave-like indentations blasted out of the rock, and the men tend to be fairly comfortable except for the noise. The raise borer is moved about on rails; it can be reduced to about seven feet to fit on the cage and is on its own carrier. The men operating the machine average 20 to 25 per cent bonus. It takes about six working days to ream a 180-foot shaft and about three days to move and set up in another area.

Raise borers have been widely adopted in most of Inco's Sudbury mines and all its Thompson mines, where their use began in 1969 in Birchtree Mine. The raise borer crew and the person operating an in-the-hole drill do much the same work. The technology is similar and the tasks just about the same except that the scale of the raise borer is somewhat larger and much of the work concerns moving it from one location to another. But there is no question about the difference in skill levels between the raise crew and the crew operating the raise borer. The raise crew does virtually all its work by hand, controls a typical mining cycle of drilling, blasting, mucking, and timbering, and must know how to set the charges correctly to keep the raise straight. The raise borer drill crew in essence tends a machine and occasionally adds another length of pipe. They have little control over the pace or quality of their work, as the different bonus rates for the two types of crew indicate.

Moving Ore within Stopes

* * * * *

A Sudbury miner talking about 1940:
You went in the mine as a shoveller. Sometimes you spent a few months just cleaning up on the level; then, when they needed a man in the stopes to shovel the muck into the chutes, you would get your chance as your turn came to go into the stopes. You would be surprised how much muck those shovellers could move.

* * * * *

Traditionally, a great deal of the bull-work in mining was done at the end of a shovel. While virtually every miner still begins his career with a shovel in his hands, this phase usually lasts only a few weeks – a kind of initiation into the rigours of mining. In the early days each mining crew included several shovellers who put the ore into wheelbarrows or small rail cars from which it was unloaded down the ore passes. The slusher was a great advance. Not introduced into Canada until 1930, it was soon widely adopted. The slusher is quite a simple device. It consists of an engine that powers and regulates the movement of a large metal scraper blade which moves on two heavy cables extended between the slusher and the end of the stope or drift. By moving the blade back and forth the slusher operator, sitting at the controls of the engine, pulls the ore from the face of the stope to an ore pass. The following conversation with an old Inco miner illustrates the impact of the new machine on the organization of work in the late 1940s:

How did it come about that you moved from working with a crew of people to just working with a partner – just the two of you?

It was the development of the slusher in particular. The

change from the three- or five-man crew down to a two-man crew was brought about by the development of the slushers. When we got slushers, instead of having a chute every twenty-two feet – all the length of the stope – we were able to slush all the muck into one chute, unless it was a very long stope; then we would have to have two chutes. [In a square-set timber stope], *while one man was doing all the mucking down below, another man up on the mining floor could be cleaning up, scaling, barring some of the muck pile down, pushing the booms forward, and getting ready to boom out. By the time he got the booms all forward into the open, his partner would come up and give him a hand to put the bulkheads on the booms. Well, then one man would continue slushing while the other man would bar all the muck down. In the wide stopes they generally had three men, but in the earlier days on the one shift they would have five men because of the shovelling and chute work.*

What kind of skill does it take to operate a slusher?

It is a perfectly simple thing; you have a giant hoe weighing three, four, or five hundred pounds. The slusher has two cables, one to the scraper and another going away back to a point beyond and hooking onto the rear end of the scraper. So you pull the scraper back and forth over the muck pile and then pull the scraperful to the chute; nothing to it. But one time, my partner went on vacation, and they got another man to come in with me to take his place. This new man was putting down the same muck in an hour's time as it took my regular partner half the shift. Now it may not seem possible, but it is possible; and what you can do with those scrapers, if you know how, makes a tremendous difference.

Until the late 1960s slushers were the primary means of moving ore within production stopes. They continue in use in many mines but have been replaced in others by scooptrams, which are able to move fifteen times as much ore per man-shift. In the most recently developed mines, Levack West for example, scooptrams and other trackless diesel vehicles have completely eclipsed slushers. As indicated in the miner's description above, the slusher is a simple piece of machinery but one over which the operator has a great deal of control. A skilled slusher operator can move many times more ore than a novice (or even an old-timer without the knack).

Scooptram operators, on the other hand, have much less control over the productivity of their machines. In fact, the greatest control they can exert is to keep the machine operating by preventing its breaking down. In a very short time (a few weeks at most), they can learn to have the machine operating at its capacity. Inco's first underground diesel vehicle was introduced in March 1966 at Frood Mine; it was a 145 horsepower scooptram (called the ST; see Figure 17). These four-wheel-drive diesel-powered units are like front-end loaders. Their buckets hold from eight to twelve tons of ore, and the vehicles themselves are from seventeen to twenty-nine feet long and five to eight feet wide, with rubber tires four feet in diameter. By 1969 there were ninety-eight LHD (load-haul-dump) machines in use at Inco mines: seventy-four with four- or five-cubic-yard buckets, two with three-yard buckets, and twenty-two with two-yard buckets; in 1972 there were 190 LHDs, the largest having an eight-cubic-yard capacity. A study of their efficiency in 1972 showed that mucking costs per ton were ninety-two cents for ST-2s, forty-nine cents for ST-4s, and thirty-one cents for ST-8s. The largest machines were able to move eighty-five tons an hour.[8] The machines are mobile and versatile but not without their problems, which are, according to the *Engineering and Mining Journal*, "diesel exhaust fumes, heat, and dust, the need for skilled operator [?] and maintenance manpower, increased capital investment, higher repair and maintenance costs for machines and roadways, and the need to stock larger amounts of spare parts."[9] But the advantages for the mining companies, if faced with labour shortages (particularly of skilled miners) and rising costs of production, far outweigh their disadvantages. It is not always a simple matter, however, to put scooptrams into stopes. Problems include ventilation, maintenance, and the amount of operating time (see pages 150-51). Some new mines, such as Levack West and the revamped Creighton No. 3, have been

Fig. 17. An ST-4 scooptram

Canadian Mining Journal 98 (May 1977)

developed specifically for scooptrams, operating with ramps from the surface and with service bays underground; in some others, ramps or inclines have been built within mines (as in Sudbury's Creighton No. 9 or between the 1,900- and 2,300-foot levels at Birchtree Mine in Thompson), or captured scoops have been put into large stopes. In ramp mines scooptrams are usually supplemented by load-haul-dump machines to move ore greater distances. The LHD machines are loaded to their twenty-ton capacity by scooptrams and haul the ore to chutes.

* * * * *

Asked about the difference between scooptrams and slushers, a Thompson miner comments:
I've worked in slusher stopes and diesel scooptram stopes, and now I'm in electric scooptram stopes. The only problem we have with the electric scooptram is it breaks down a lot, but it doesn't affect your health so bad. It's a lot more powerful than diesel; it's got more hydraulic power

that can lift the muck. When it's going it's a better machine, but it doesn't always go; it's one thing then another. They made it too light a machine for too heavy work. . . . The slusher stope was not perfect to work in, but it was pretty nice. You work in a scooptram diesel stope, and half the time when you come up you can hardly breathe; I would choke and spit. Now, I work with an electric scooptram and I choke and spit but not half as bad. No noticeable heat either, like the diesel.

* * * * *

The first scooptrams were introduced in Thompson in 1967, just after their appearance in Sudbury. The first electric machines appeared in early 1975, but they were relatively smaller, with a one-yard capacity and a fifty horsepower engine. These smaller scoops are more appropriate for captured stopes in Thompson, where machines are not worked as hard or as constantly as in the "go-go" mechanized areas in Sudbury, where only mobile equipment is used and is in constant use. Within three years of their introduction in Thompson there were forty-seven scooptrams accounting for 62 per cent of the daily movement of ore. In Sudbury, where some completely mechanized mines exist, it is likely that scooptrams account for an even greater proportion.

The mines with the greatest degree of mechanization in the movement of ore are, of course, the open-pit mines (such as Pipe Mine, discussed earlier), where much of this technology was originally tried. The giant shovels and haulage trucks have not been adapted to underground mines, but many of the drills and Caterpillar dozers have already found their way into these operations. As the scale of underground mining increases there will be more adoption of this open-pit technology.

Moving Ore in the Mines

There have been other changes in underground mechanization but none so impressive as in-the-hole drills, jumbo drills, raise borers, and scooptrams. Underground, after the ore has been removed from production stopes it is still moved mainly by trams.* The old underground rail systems had small twenty-cubic-foot cars moved by hand or by horses. The first electrified underground equipment, locomotives, appeared around 1910 and were followed soon by battery locomotives. Diesel locomotives were introduced in 1950, and today both types are used to move ore. There has been some experimentation with conveyor belt systems, but they have not been widely adopted.

Tram crews usually consist of two men, a motorman and a switchman. Their task is to move ore from various draw points or chutes to larger ore passes. The switchman walks in front of the tram as a safety measure and switches tracks if necessary while the motorman drives the tram. Often each man is qualified for both jobs and they alternate tasks. In some older mines, as at Crean Hill, the switchman is also a slusher operator, but in most mines he only operates chutes for loading and unloading the cars. A member of a Thompson tramming crew said of this operation, "You have to be very careful pulling the chute. You don't let just anybody do it. But once you get to know your partner and you work together every day, you can just about read his mind. Pulling the chute will make you sweat if you have to bar for a considerable period of time. On the other hand, you may have to get some powder and blast. Again, that's not always successful. You may blast two, three, four, or five times and there may be no movement." Hang-ups of ore are the main impediment to moving the ore quickly. Besides these, there are other problems, as another Thompson motorman tells it: "You've got to clean up the messes when the muck gets on the track, and you have to put your train back on the track when it comes off . . . it all depends where you are. Some tracks are good, others terrible. I spend 50 per cent of my time putting my train back on the track." Both hang-ups and derailed cars are dangerous and the source of most injuries. The trams also create dangers for other miners, as will be discussed in Chapter 7.

Although tramming has not experienced the major change

* The exception is ramp mining (Creighton No. 3, Levack West, and part of Birchtree), where ore is loaded into load-haul-dump machines by scooptrams and transported in these rubber-tired vehicles to ore passes (see page 109).

that has taken place in the drifts, advanced technology has been introduced into the larger, longer ore haulage areas. This is the case at Levack West, where ore is hauled to the old Levack Mine for movement to the surface, and in Thompson Mine between the T-3 and T-1 areas. At Levack West, ore is moved by a twenty-five-ton trolley locomotive with twelve cars run by two men, one driving the train and the other operating the ore chute. These men are on an individual bonus at Levack Mine and with another crew work on a two-shift basis. The cars are loaded from a chute much like that used by the drift trams but larger in scale, controlled by the operator, and the operator is responsible for cleaning up spillage. A similar bottom-dump car system used in Thompson Mine is illustrated in Figure 18. This is somewhat larger, with a diesel locomotive pulling thirty cars, and must haul the ore a distance of two miles underground to a holding bin, with the round trip taking about an hour. Derailment, ore pass hang-ups, and spillage create problems for these trains as with the smaller trams but not as frequently. The Thompson equipment, costing about $2 million to introduce, replaced a system in which the cars were emptied by being tipped on their sides, which tended to cause the ore to jam in the cars and build up in the chutes.

Ore from the holding bins is then passed to the crusher areas, which perform the first stage of milling the ore. It is only in this part of the mines that any automation has appeared. The ore is first crushed by swing hammers, then moved by a vibrating feeder into a huge crusher, and finally moved by conveyor belts to the loading area, where it is hoisted to the surface on skips. The conveyor system has automatic controls, as does the crusher. Men at control panels oversee the process using television monitors. The cost of this system was about $1 million in the Thompson Mine.

* * * * *

A Sudbury miner describes traditional mine transport: *The trains went through a main ore pass, and this ore pass dropped the ore down to a big crusher underground; the muck would go through the crusher. Well, then about one hundred feet farther down there would be a fill pocket to load the skips to carry the ore to the surface*

Fig. 18. Bottom-dump cars unloading muck into an ore pass

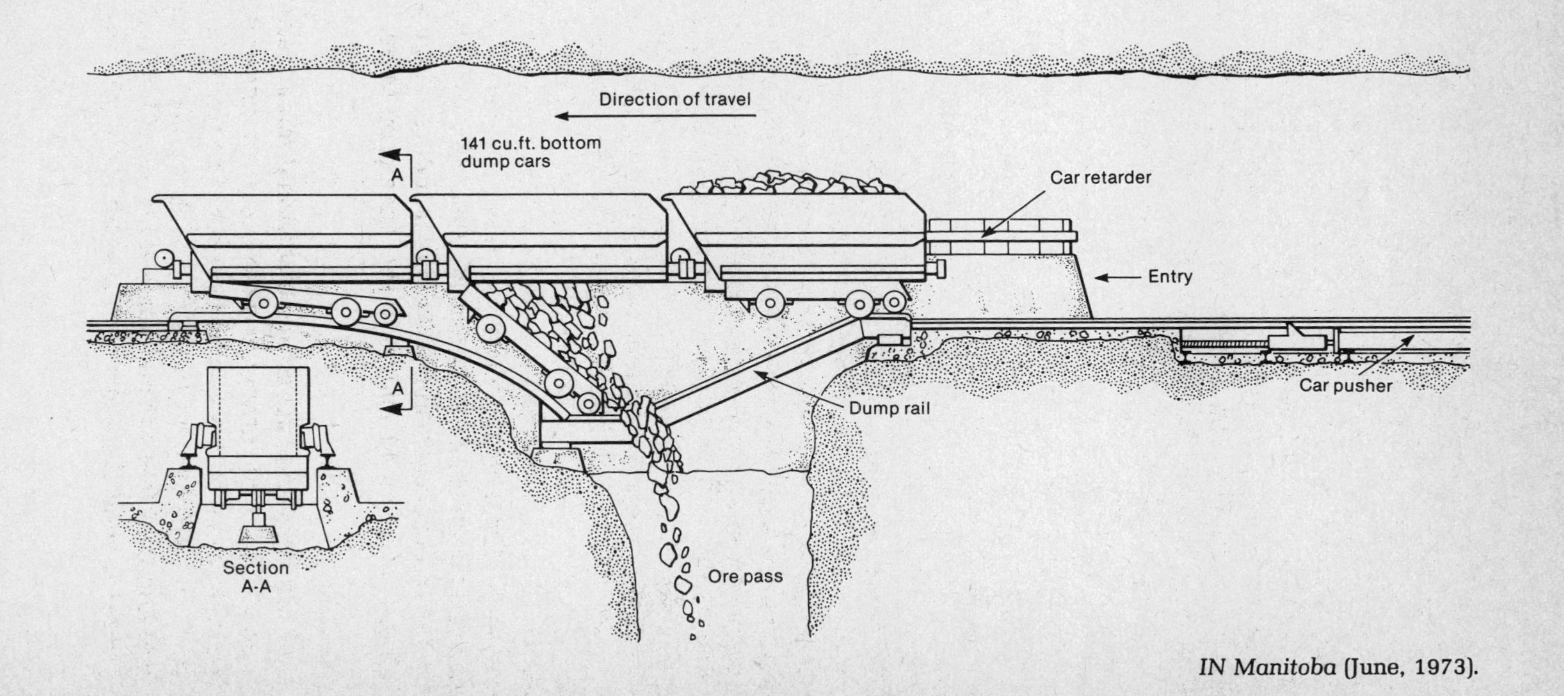

IN Manitoba (June, 1973).

(the skip's the shaft where all the supplies were brought down and all the ore was hoisted). It was hoisted up into the rock house, and there was another crusher there to make the muck a little smaller. From there it went up to the smelter where they had a big cone crusher to crush it down to about three-quarters of an inch.

* * * * *

The farther it is away from the production stopes and development work in the mine, the more standardized and subject to mechanization the work becomes. Mechanization began earlier and has gone farther in the transportation of ore within the mines than in the production and development areas. But the first signs of automation have also made themselves evident in the last stage of the underground operation. It can be anticipated that automation, which reduces the need for labour even more than mechanization, will gradually work back into the mining process. If this happens, the next area for automation will be the long-haul tramming followed by the short-haul tramming, and the obvious way to do this is to use a conveyor system. To date it has not been possible to use such a system because the pieces of ore are too large, but this does not appear to be an insurmountable barrier. Conveyors would have several advantages, as the *Engineering and Mining Journal* points out: "Continous material flow, low operating cost, low personnel requirements, and elimination of locomotive maintenance."[10] It is apparent that in the near future conveyors will be used more extensively, and they are likely to be connected to television monitoring systems and computerized loading and metal detection systems, techniques that have already found their way into underground crushing systems.

Work is even being done on "non-entry" mines, in which automatic machinery will take the place of underground workers; examples are "auger mining, underground gasification, or combustion of coal deposits, *in-situ* leaching and solution mining."[11] The idea behind *in-situ* mining is to dissolve metals with chemical solutions and bring the liquid to surface as a slurry (a thick liquid). There are no such mines in Canada, and there is no likelihood that these methods will be practised in the near future; but they remain possibilities.

Although automation has not been a major force underground until now, there is some evidence that in the next years its impact will be felt. "Integration" of mining is foreseen by the *Canadian Mining Journal*: "Initially computer programs for calculating ore reserves were mainly concerned with long range planning rather than with materials movement and operations control. These programs have since been extended so that alternatives in haulage methods and equipment could be studied. It would appear these simulated techniques are moving closer to real time operations control, treating the entire mining operation as a total integrated system, from ore face to stock pile or concentrating plant. If true, this would have considerable significance."[12] The possibility exists for computerized drilling, blast-hole loading, conveyors, pumping, ventilation systems, and various other operations. But adoption of these technologies, even if they were developed, would not be a simple matter. The existing mines (and surface plants, for that matter) impose limits on what can be done. It is only in the newly opened operations that all of the latest technologies can be applied. This is evident in Levack West Mine underground and in Clarabelle Mill and the Copper Cliff Nickel Refinery on surface.

Maintaining Underground Equipment

Mechanization clearly saves Inco a great deal in labour costs but not without incurring many other costs. The current range in the cost of labour as a proportion of production in Inco mines is from less than 40 per cent in the most mechanized to over 70 per cent in the least mechanized mines. The capital outlay for the equipment is tremendous, but a less obvious cost with mechanization is the outlay for maintenance. The following conversation with a senior Inco engineer in Sudbury illustrates the scope of the problem:

> *Our big cost is maintenance. It is not unusual to spend in a year 100 per cent of your capital costs mainly on maintenance . . . and sometimes it can be a couple of*

times that. If you go and buy the parts to rebuild a machine, it will cost you about four or five times the original price of the thing. If you buy a drill by pieces, it is about three times the price of the drill. A great deal of the suppliers' money is in replacement of parts and, of course, then you try to find other ways to get them a little cheaper here or there, or maybe even make them. It really ends up with a very high maintenance level, so the maintenance people get very much involved in who is going to buy what. Whether you spend $150,000 or $200,000 for a machine is not nearly so important as how much it is going to cost to keep it going. For example, the cost of recapping a tire is $6,000 or $7,000; just insane, the prices. And you have to pay to get somebody with a knowledge of tires. Of course, the worst thing that can happen to tires is water. Let's say there is water in a hole, and if there is a sharp rock in it, it will very likely cut the tire. Water is used as a lubricant for cutting rubber; so puddles and where water goes is important, and your drainage system gets very, very important, again on a cost basis.

Is it very hard to maintain equipment underground?

It is harder. On surface water will drain, but there you don't have any place for it to drain because there is solid rock underneath. So your road beds won't drain like normal road beds.

Conditions underground, particularly ventilation and drainage, have a strong bearing on the possible use of trackless diesel equipment. A year and a half elapses between a decision to put a new ventilation system in an existing mine and implementation of the project. Diesel equipment necessitates greater ventilation because it produces gas fumes that must be removed from the drifts and stopes, and many of the older mines are simply not constructed to accept any more ventilation without large development costs. Obviously there are also important health considerations (see Chapter 7).

Working in a mine poses particular problems for maintenance crews. In the large ramp mines they have been overcome to some extent by construction of large underground garages (see Figure 5), but most of the ore moving equipment is located in captured equipment stopes. The initial problem is to get the scoop into the stopes. It must be disassembled on surface into pieces that are a maximum of seven feet long, taken by cage to the stopes, and reassembled. This process takes a pair of maintenance workers about two weeks. All repairs must then be done in the stopes. For parts, the maintenance man must call the surface warehouse. It then takes a minimum of half an hour and sometimes up to three hours for even a nut and bolt to come down from surface. The job may be minor, such as repairing a hose on a scoop, but can include overhauling an entire engine. The ventilation is often poor, and the maintenance men are always working on sand or rough ore. When repairs are required it takes double the amount of time to do them in the stope that it would in a shop. Even setting up a hoist requires drilling bolts into the walls to hold the chains; all the work must be done by hand, and the only light is that from the helmets.

The machines themselves have heavy demands on them. A Thompson millwright who has been on maintenance work for four years says he has four machines on his level, and often three will be broken down in the morning. He argues that much of the maintenance work is caused by production crews pushing their machines too hard because they are "bonus hungry."

* * * * *

A Sudbury miner on maintenance:
Why would maintenance be poor? Is it not in the company's interest to have things fixed when minor problems arise rather than have them break down?

No doubt they would agree – the big guys. They don't have the pressure on them like shift bosses. We've seen shift bosses being screamed at in the office, literally screamed at, by their general foremen. They get threatened probably two or three times a week if the muck's not going out. And they react, and the way they react is produce.

* * * * *

Implementing Technological Changes in Mining

Introduction of new methods and technologies in mining is governed by important constraints. There are more of these than just the sources of capital and the market conditions discussed earlier and even the development of new techniques. Some are physical: the nature of the ore body, the method used to develop the ore body in the past, and the capacity of the various services in the mine. A particular technology is not automatically implemented just because it exists. Its installation is likely to be sporadic and intermittent at the discretion of management. It may be adopted on a large scale in new mines or when development work can be done to adapt old mines. In many cases only parts of the new technology can be used. The form of mine entry, which is guided by the depth of the ore body, relates to the kind of techniques adopted. Open-pit entry has a limited depth of less than a thousand feet, while ramp mines can go deeper, perhaps two thousand feet; shaft mines are still viable at over seven thousand feet.

There is little doubt that sometime in the future most mines will be mechanized but for the present, as Ronald Taylor, president of Inco's Ontario division, said in the *Canadian Mining Journal*, "I can't say we've got an overall plan that we'll completely mechanize every mine. You don't want to put a Cadillac in where you only need a Model T Ford."[13] The following overview is derived mainly from an interview with a senior Inco engineer:

> *We have tried to improve our productivity a great deal with mechanization, and that isn't very fast or simple. It generally takes you in the order of eight years from the time it's decided to spend money on any major operations before you produce something. To sink a shaft, you have to open the area. You might do it a little faster, but you'll never do it under five years. What you are talking about basically is an eight-year lead time to make your plans, gather your money together, start to develop, and actually get into production.*
>
> Do you have to decide in great detail the actual methods of mining, that is, if you are going to use the large diesel equipment?

Your ventilation, which you have to start off with initially, has a big cost in it and is a big controlling feature about how you can mine and what type of equipment you can use, and this is a big part of our problem going to mechanized equipment. You can put the equipment in, but you don't have the ventilation. Or, if you want to put another ventilation system in an average mine you are talking about a year and a half. If you decide to do it, you get another piece of ventilation there and you are talking of a minimum of $1.5 million for the separate kind of ventilation system, but it can be up to $12 or $15 million for some of the deeper mines for a basic ventilation addition.

It is very difficult to put in the kind of capital money [an old mine] *deserves to start fresh, and often you don't have any other locations left to put holes or shafts or drifts in to get to the varying locations. The initial planning of how you are going to do it and what you are going to put in is pretty darn important, but greater investment in old mines is also rather expensive over the years for a good return.*

There has been first a levelling off and then a decline in the total number of people working for Inco in Canada, yet many operations have expanded and new ones have opened. The company continues to open new mines and sink new shafts. It appears that technology has been responsible for holding the entire labour force at or below the earlier level. What kind of impact does that have on the company? A senior Inco official responded:

Well, we are still very labour intensive. Again, I am not very sure how we relate to other companies, but in the older mines we are talking about, something in the order of maybe 70 per cent of your total costs are labour costs. In the new ones that we have been able to design to fit some of the present technology and equipment, we might be down to 40 per cent or even less of our total cost in labour.

What type of labour problems has the company experienced with regard to technological innovations?

I don't think very many. The younger and newer

employees like to drive some of this mechanized loading and hauling equipment much more than the older employees. Generally speaking, the fellow goes in the mine and he's there for the money – most everybody is – and he wants the most he can get the fastest. He stays with the company for quite a long time. He might work for about the first ten or twelve years on jobs that are incentive related. Then after that maybe he has got his house and his camp and so on, and he starts to look for a little bit more sedentary type job to bid on and generally will get out of the incentive work. Or he might move on to the tram area or some other type of work, so you'll find that you have different age levels and groups working at different types of mining operations. One of the older methods is an undercut-and-fill method for recovery of pillars, which was invented about fifteen years ago. It turns out to be a rather high-incentive paying job, and then you will find very senior employees.

One of the things that is a real paradox is the way you speak of miners as very skilled persons who are able to drill and set and blast and command umpteen dozen skills, yet there seems to be a move away from that kind of mining and into mechanized mining, where you are operating a lot of equipment. Are those people increasing their skills or decreasing their skills?

They are getting to be quite a bit more mechanized, and this requires knowledge of machinery, hydraulic hoses, care of equipment and what people do with it. It certainly has changed. I would say that basically they are increasing their skills.

Do the workers who operate the machinery also do the maintenance of the machine?

No. They will do very little. They will put in the oil and check the hydraulic tank and change the tires, and they might change the hydraulic hoses, but basically, that is about as far as the operator will go. Then you have the diesel mechanic basically doing the rest.

It is interesting to note that this senior engineer's immediate reaction to the question of whether mechanization increases or decreases the skills of the miner was that they are increased. But when pressed, he acknowledged that much of what he regarded as skilled work was actually being performed by maintenance workers, not the operators of the equipment. Finally, he stated, in another context, that "the more automation and the more refined the equipment, the less the particular worker can influence the job; it is the piece of equipment that sets the pace, and a lot of it is how he uses it. If he breaks it by misuse, it costs him productivity and money, if he is on the incentive system." With sophisticated mechanization, the worker is not being paid for how hard he works or how skilful he is; he is being paid to keep "his" equipment clean and functioning. Although he may not have any great influence on how quickly the machine runs, he can have an effect on its down-time, through either neglect or conscious action. The earlier contrast between shovellers, slusher operators, and scooptram operators is a case in point. The first obviously directly controls the pace of his work, the second can use his skills for a substantial effect on the rate of his output, but the third is essentially governed by the capacity of the machine.

Nearly one-tenth of the workers in mining are in science-related fields (see Appendix V), and most of these are engineers. At Inco, what do most of the engineers associated with the mines do? This was the answer of a senior Inco engineer:

> *Well, when you have an ore body, you have various* [mining] *locations within that particular mine and different grades, and you will have to plan in fair detail for at least five years. I am not talking about developing a mine. In an ongoing mine you will have to detail for at least the next two years exactly what hole and what drift and where you have to go and how it is going to get involved. You will have a pretty fair plan for at least twenty years down the road. You have to decide if you are going to spend umpteen hundred million bucks and*

whether you are going to get it back. So the actual mine planning in that range takes a fair bit of work. Then you get down to the detail of developing and opening up a particular stope or working place or small locale. It takes quite a bit of detail to place it so there is no interference because everything is piled on something else. It is like having a one-way street. Whatever some fellow does there affects something else, and your hoisting capacity and your pumping capacity as well as your ventilation and where all the services fit has to be planned and taken into account in every single thing you do.

Engineers spend a good deal of their time working underground in conditions similar to those of the miners. Some spend half of their day with the shift bosses and miners determining specific mining techniques. They are constantly measuring the stopes and, as will be seen later, determining the amount of bonus to be paid. The mining geologist, whose drawings are the raw material from which mine engineers plan their layouts for drifts and ore passes, also spends a good part of his time underground doing testing. According to *Inco Triangle*, "The mine geologist spends about half his time underground, estimating grades and marking the limits of the ore at working stopes, providing an 'ore map' for the miners to follow."[14] Thus workers in these "middle-class" occupations are often subject to the same working conditions as the miners themselves. The two groups appear to have a fairly good working relationship because the engineers and geologists are not immediately engaged in directing the miners and there is usually solidarity among all who work underground, regardless of their task.

The Organization of Underground Work

The nature of supervision in the mines is ambiguous in a number of senses. It is, on the one hand, traditionally very tough; as will be seen in later chapters, supervisors have historically exercised very arbitrary and at times ruthless power over workers. On the other hand, miners have had a great deal of autonomy and even now seldom see their supervisors, on the whole organizing and pacing their own work. Besides this basic ambiguity there is a tendency for the nature of supervision to change as mechanization increases and work areas grow larger. As work in the mines is removed from many small production stopes and centralized in a few large areas, supervisors can keep a closer watch over workers and workers themselves have less discretion in organizing their work because they tend to be confined to the operation of one machine and one task.

Supervisors (or shift bosses, as they are called by the miners) are themselves in an ambivalent position in the hierarchy of the mines. They are between the firing lines of workers and management. In all of the shaft mines they are expected to cover, on foot, a very large area with many distinct work places; in ramp mines they have access to vehicles that can move them quickly from one area to another, and a lower number of workers are concentrated into fewer work sites. Yet shaft mines are still in the majority at Inco, and although the supervisor is pressured by management to insure production, he cannot directly oversee the men's work.

Management selects supervisors from their own list of "designated individuals" who are trained in various aspects of mining. Often these are miners who have had more schooling than the average or have distinguished themselves through union work. Union stewards, for example, are often asked if they would like to become supervisors. Supervisors in training report periodically on their progress to the mine superintendent. These visits appear to be more for socializing than for actual training.

Supervisors are responsible for workers in a section of a mine, usually about thirty men but ranging as high as fifty. Although hourly rated workers have eight-hour shifts, supervisors normally work nine-hour shifts, arriving three-quarters of an hour before cage time to check in with the engineering office and do administrative work. At the end of shift they record what has happened on their shift: production, accidents, penalties, and problems. This takes up to half an hour daily. The rest of their shift they spend underground with the miners, going down in the cage with them and eating lunch with them. Each day the supervisor allocates tasks to the men on his shift, keeping a log of activities from beginning to end of the shift. These allocations of work are the supervisor's interpretation of the directions of the management and engineering people who draft plans and schedules. He decides when there are men available to do specific work and how much will be done and evaluates the quality of the work.

Supervisors report directly to a general mine foreman who usually oversees about six supervisors. Seldom do they deal directly with the superintendent, who is the highest-ranking person at the mine site. These general foremen take part in planning and work more closely with the engineers than do the supervisors. Supervisors are evaluated in terms of their production and problems. As a supervisor of seven years at Sudbury puts it, "It's a numbers game, I would say. The amount of dressings [injuries], the amount of tons of ore that you produce."

From the workers' perspective, it is the shift bosses who regulate the pressure to produce. They are the company overseers and watch-dogs. They are the intermediaries between capital and labour. A perceptive Sudbury miner said of their reaction to less ore output than expected, "I've heard it a hundred times, 'You're doing it to screw me. You are just doing it to screw me or to screw the company, not because it's unsafe.' It's a syndrome they have; the Shift-Boss Syndrome; everybody is against them." The shift boss is in no-man's land, as this miner went on to say: "It's a rotten job. But everyone figures they're on the way up. They don't figure they are going to stay there, but 99 per cent are going to stay. They can't go either up or down." Supervisors usually exercise control not by directly overseeing the work process but by regulating the amount of work done with the aid of the bonus system (see Chapter 8), which is also based on performance. As a Thompson miner says, "Nobody is breathing over your shoulder, except the bonus."

Not all regulation of work is done by the supervisor. The workers themselves are arranged hierarchically with regard to accountability, and they regard this system as their internal organization of work. All hourly paid workers are supervised to some degree, but those in one group are supervised directly by salaried supervisors and do not direct others, those in a second are supervised but themselves direct other hourly paid workers, and those in a third are supervised by salaried supervisors and directed by hourly paid workers.

The following are some of the main underground job classifications corresponding to these different forms of supervision and direction:

Supervised by shift boss and does not exercise direction	Supervised by shift boss and directs others (hourly paid)	Supervised by shift boss and directed by hourly paid workers
loader operator	raise driller	driller
fuel supply man	drift leader	drift driller
bulldozer operator	stope leader	LHD operator
trammer	pump man	track man
switchman	sandfill boss	cage tender
mine beginner	skip tender	diamond drill helper
powder man	timber man	raise drill helper
motor crew	track boss	nipper
motorman	top man	sandfill helper
labourer	lift truck operator	timber man helper
hoist man	fitter general	scrap picker

It is apparent that the "teams" of miners themselves have clearly defined relationships whereby some miners give direction to others. All the classifications in the middle column give direction to workers in the third column while those in the first column operate directly under the supervision of the shift boss.

One of the best examples of an hourly paid worker directing others is the stope boss. This person is similar to the stope leader in the more conventional forms of mining, except that the latter actually works along with a driller, but the stope boss directs more people and does not work physically. One stope boss explained, "I don't work, physically; I just tell people what to do. I have twelve men in one of those 'go-go' areas." (A go-go area is one that is highly mechanized, where machines are constantly operating, as opposed to a conventional area where there is a crew of perhaps two men.)

* * * * *

A young Sudbury miner:
No matter what job you are doing, there's very, very little supervision. Why, you look at the mine in total perspective; how can one man be in all those different places? It's actually a city under there. A level's a big place. Some are over a mile from one end to the other. A man can't cover that much ground. Each of the workplaces is completely separate from the other, as far as the stopes go. In order for a shift boss to be there, you would have to have a shift boss for each individual stope, which is not economical. What takes the place of supervision is that if you don't get the job done, they know. You have to report in each night. Your shift boss is there once a day anyway, so if you told him, "I've done this and this," and he walks in and you're just doing it today, and you told him you did it yesterday, it's obvious he knows you weren't doing it. There's that kind of check. It's not direct supervision.

* * * * *

With the centralization of production in the mines through mechanization, supervision becomes more direct because there are fewer production areas and the shift boss can cover them more readily. Moreover, there is more differentiation among the miners themselves. Not only do they become specialized in performing one task but the stope leader, who used to work alongside his driller in a partnership, now becomes a stope boss, who directs other men instead of performing the work.

These developments in hard-rock mining parallel much earlier changes in coal mining* in Britain, reported in a classic article by E. Trist and K. Bamforth entitled, "Some Social and Psychological Consequences of the Longwall Method of Coal Getting." Traditionally, coal mining was carried out by "interdependent working pairs" who were responsible for the complete mining operation and conducted the entire cycle of

* Coal mining is soft-rock mining and thus more readily subject to mechanization.

work. Trist and Bamforth argued that "leadership and 'supervision' were internal to the group, which had a quality of *responsible autonomy*. The capacity of these groups for self-regulation was a function of the wholeness of their work task."[15] When the mining method was centred on "short walls" and the miner and his partner formed the production unit, "his equipment was simple, his tasks were multiple. . . . He had craft pride and artisan independence." There were no "intermediate structures" overseeing the direct production process.[16] However, "mechanization made possible the working of a single long face in place of a series of short faces . . . bringing into existence a work relationship structure radically different from that associated with hand-got procedures. . . . A structure of intermediate social magnitude began therefore to emerge."[17] This same force has been in operation in the Inco mines since the late 1960s and is currently reorganizing the nature of work and the structure of control in the mines. Management now has more direct control over the work process, workers themselves are being ordered in a hierarchical structure, and there is a clear tendency toward the loss of craftsmanship in the art of mining.

It is not simply nostalgia to lament the passing of the skilled miner; for the miners themselves, the skilled trade of mining has been a way of life that has given them pride. They are craftsmen with finely developed talents. Yet the work is hard and dangerous. The elimination of bull-work is welcomed as much by the miners as by anyone else. Yet gone along with the bull-work is the independence and creativity of the job. There has been a general levelling of skills by eliminating both the most and the least skilled. Gone is the organization of production by "teams" of workers who, in spite of working for wages and in spite of being subject to company regulation, managed to a great extent to retain their autonomy and control over an entire work cycle. Although this is the result of the introduction of mechanization, it is not the inevitable outcome. The way in which technology has been introduced – the interests it serves – has been controlled by capital, not by labour. Work has been reorganized *for* the miners, not *by* them. With the increase in mechanization it has been possible for management to penetrate, to an extent they were unable to in the past, the miner's control over the pace of his work and the skills he brings to bear. Management's strategy in introducing technology has been to decrease its reliance on the skills of the miners and to minimize the number and quality of workers needed. The goal is to increase the company's control over the work process and maximize profits. As will be seen in subsequent chapters, the miners themselves have lost in many ways – in their ability to demand bonus, in the depreciation of their mining skills, and all too often in their health and safety. Technology is not neutral in the relationship between capital and labour because it is designed from the outset to meet the needs of capital, not labour. Nevertheless, it does have the potential to humanize the labour process if it is adapted to be most beneficial to those directly affected – the miners.

Obviously miners have not accepted these developments without resistance. In an organized way they have fought the health and safety issues and, at least to some extent, appear to have made some gains. They have also, through their unions, been fighting for their position with programs such as "miner-as-a-trade" (see page 341). For the most part, however, they have accepted the technology as inevitable. Many of the old miners simply continue in their trade, and younger miners adapt easily to the mechanized equipment because they can learn to use it quickly and do not need to spend years gaining experience. All too often the introduction of technology is accepted as management's right simply because management claims it is. The unions have not been very alert to these developments; they have done little systematic analysis to understand their implications. Although their ranks are dwindling, they see few ways to overcome the problem. Watching their members lose their autonomy and the company appropriate greater control over the workers, they still too often continue to accept a managerial prerogative in how technology is introduced.

CHAPTER SIX

From Traditional to Ultramodern: Inco's Surface Operations

Traditionally, surface work in the mining industry has been a combination of bull-work and craft production; the operations have been labour intensive, even though there has been a high degree of mechanization, because they required workers to perform a great deal of detail labour. There has, however, been a strong tendency on surface, particularly evident since the early 1970s, to move from mechanized production to automated production. This is most obvious in the newly constructed facilities such as Clarabelle Mill and the Copper Cliff Nickel Refinery, but is also evident in alterations to older plants. The consequences for labour are just as significant as those caused by the changes underground, although obviously of a different sort. Much of the bull-work has been eliminated, but so has much of the craft production; they have been replaced in many instances by dial watching and patrol duty, making sure that the automatic equipment continues to function. Many surface workers in the automated plants no longer have routine tasks. Instead, they monitor equipment and make repairs when necessary.

These changes are by no means universal, but the tendency is clear. For each major processing phase at Inco there are plants with contrasting methods of production that allow an analysis of the direction of change in the nature of surface work in the mining industry. In this chapter the surface operations of Inco at three mills, Copper Cliff (1930), Thompson

(1961), and Clarabelle (1971); the two smelters, Copper Cliff (1930) and Thompson (1961); the three nickel refineries, Port Colborne (1918), Thompson (1961), and Copper Cliff (1973); and the only copper refinery, the Copper Cliff Copper Refinery (1929, with major changes), will be examined with reference to types of technology and some specific changes in the numbers and types of workers required. Appendix IX indicates the products that flow from each of the smelters and refineries and their destinations.

When people in the mining industry talk about the difference between surface and underground work, they often say jokingly, "It's like night and day." As far as it goes, this is true. Work underground is radically different from that on surface. On the other hand, both are subject to some common forces. The most important of these is capitalization. It is expressed underground as mechanization and on surface as automation, but in both cases the consequences are similar. Capitalization allows management to use science to penetrate and control the work process, thus reducing the demand for labour. Production per employee in two nickel refineries illustrates the difference. In 1976 about 60,000 pounds of nickel were produced for each employee at Port Colborne (PCNR) while at Copper Cliff (CCNR) the figure was 360,000 pounds, or about six times as much. With one-quarter the number of employees, CCNR produced about 50 per cent more nickel than PCNR. In 1978 PCNR was moth-balled and CCNR took its place. Automation reduces the demand for skilled labour on surface but increases it for maintenance work, much as mechanization does in the mines.

Because mechanization was put into effect on surface much earlier than underground, supervision there has always been much more immediate. The workers are located in centralized operations. With automation, however, the nature of supervision again changes because there are fewer workers, spread out over a broader area. Management needs workers not to perform constant operations but mainly to service equipment, watch for problems, and be available for maintenance. The performance of this work is ensured by instrumentation monitoring the equipment and through radio contact with workers. The most important general impact is the reduction of the number of workers required for production, but it is also felt in different forms of supervision and training for the remaining workers.

The Mills

Milling is the first process in extracting metal from the ore after it has been mined. We saw in Chapter 5 that preliminary crushing of the ore (to about six inches) is done underground before it is hoisted to the surface. Once on surface it is transported to one of Inco's five mills, three of which will be examined here. They range in condition from decrepit to ultramodern.

All the mills perform the same task: the breaking of ore into fine particles that can be separated for further processing. The steps in milling are first crushing the ore to a diameter of a half inch, then grinding the result to $^1/_{200}$ of an inch, mixing it with water (forming a slurry), and adding four chemical reagents in a process known as flotation in which the valuable minerals are separated from waste rock (or "tailings"). Rock sinks to the bottom of the flotation tanks while the minerals, attached to the reagents (now called a concentrate), are made to flow over the sides by a bubbling action caused by air introduced at the bottom of the tanks. The concentrate is then dried and sent on to the smelters.

As in the mines, the job classifications of hourly paid workers in the mills can be distinguished by the nature of their supervision. The following jobs are some of the main ones found in the mills:

Supervised by shift boss and does not exercise direction	Supervised by shift boss and directs others (hourly paid)	Supervised by shift boss and directed by hourly paid workers
pump man	grinding mill operator	grinding mill operator helper
labourer	flotation operator	flotation operator helper
crane man	crushing plant operator	crushing plant operator helper
mobile equipment operator		conveyor man

The three mills provide excellent examples of various levels of technology. Clarabelle, which is automated, produces about 125 tons a day per employee; Thompson, which exemplifies milling by instrumentation without automation, produces just under 100 tons a day per employee; and Copper Cliff, where much of the work is hand regulated, produces only 40 tons a day per employee. Their relative employee output is indicative of their respective degrees of automation, instrumentation, and mechanization.

Copper Cliff Mill

Copper Cliff Mill was opened in 1930, has a capacity of 14,000 tons a day (less than half that of Clarabelle), and employs 322 people (about one hundred more than Clarabelle). It is difficult to move around in the mill because the passageways are crowded with old equipment. The building is obviously old; much of it is damp, dark, and dirty with a strong musty smell, particularly the lowest level of the plant, which is poorly lit and reminds one of the mines. In this area three pump men and a foreman work. The working part of the main floor (one large section is closed down, its task of crushing now being performed by Clarabelle) contains banks of seventy flotation tanks, each with six sections, used to separate the concentrate and tailings. On the next level there are rows of drying equipment for turning the wet concentrate into a dry powder for the smelter. Generally the operation is a type of process, but supplementary parts are done in batches. Many of the operations are manual, such as regulating the levels of the flotation tanks and adding reagents. In these tasks the operators to some extent directly control the milling process much as they have since 1930, but there have been some significant changes in sampling procedures.

Of the 322 employees, 157 are operators, with twenty-three supervisors, 131 are on maintenance, with nine supervisors, and two are senior supervisors. At Copper Cliff, unlike Clarabelle, more men work as operators than on maintenance, reflecting the lower level of automation. The milling process has by its nature never been very labour intensive. From the outset it was highly mechanized. Now some automated equipment has been added, and automation is gradually taking place in the plant. Since 1975 the sampling has been automated. Samplers used to gather and analyse batches of slurry manually; now results of analyses appear as a computerized print-out in a small room constructed on the top floor. Thus sixteen workers have been eliminated by reducing the number of samplers from six to two per shift. The cost of the automatic sampler ($400,000) was met by the savings on labour in about *one* year. According to the plant management, much of the rejuvenation work is being done with "wooden dollars" from the internal resources of the mill by using workers' "free time" from their normal tasks. Other changes such as a simplified method of moving feed along conveyors and new driers have reduced the amount of labour required. Most of the changes have been directed at the dirtiest jobs in the plant. Other dirty jobs remain, such as that of the conveyor man guiding powder stored in stockpiles back into the process, but they are on the agenda to mechanize. Generally an attempt has been made to work within the limitations imposed by an old structure and streamline and simplify the process to the extent possible.

Thompson Mill

This mill is highly mechanized, although only a few operations are automated. The ratio of operators to maintenance workers approaches one to one. In July, 1977, there were fifty-five operators and forty-five maintenance men with additional maintenance support from the shops (besides the maintenance assigned to each plant, there is a separate maintenance department in Thompson). In total, operators had worked 50,000 hours and maintenance 44,000 hours to that point in 1977. There are eleven staff persons and five maintenance supervisors on salary in the mill. The total labour requirements in this mill have been decreasing somewhat. In 1970 the mill produced twenty-three tons per man hour. In 1977 this had increased to thirty-five tons, even though the overall output declined from 3.9 to 3.0 million tons per year. Between 1975 and 1977 the cost of labour per ton of ore decreased from twenty-nine cents to twenty-seven cents, and maintenance is now at a steady level in cost per ton.

A group of thirteen people from the process technology department work in the mill. They report to their own department rather than to the mill superintendent. At Thompson, sampling and testing at various stages in the milling is done mechanically after the material has been gathered manually. A test takes four hours, two to collect material and two to analyse it.

The mill has three parallel crushers. Once crushed, the fine ore is moved to bins and mixed with water to form a slurry for the nine rod and ball mills. In these mills the ore is crushed by being smashed first with heavy steel rods and then with heavy steel balls which move inside the mill as it is rotated. The slurry is then moved to a series of 506 flotation tanks that separate the mineral concentrate from the waste rock. About ten thousand tons of ore pass through the mill each day.

Unlike Clarabelle, Thompson does not have a central automated control room. Instead, an operator at a control panel located near the mills tells the plant workers when adjustments are to be made. Recently an automatic reagent adder was installed. Previously, one man on each of the four shifts had added the reagents, moving the chemicals on pallets and pouring one hundred bags a shift by hand. The process has now moved from "art to science," as the mill superintendent puts it. An automatic flotation system has also been added. The operator sets the desired concentration on the control panel and the levels of the flotation tanks are automatically adjusted, eliminating the task of manually raising and lowering levels (as is done at Copper Cliff).

The mill has four crews. In each one there are three men on the crusher: a panel operator, a helper, and a conveyor man. The crusher operator watches the dials and regulates the flow of "feed" so that enough ore is coming into the mill. He also checks the density of the flotation and generally monitors the process. The dirtiest job is the helper's. He is a labourer who shovels the ore that falls off the conveyor; it is heavy, hot, and dirty work. The conveyor itself is automated and runs by remote control. The conveyor man's responsibilities cover from six to fourteen conveyors, and his main job is to keep the belts clear of build-up and look after the rollers. This man must wear a dust mask and ear protectors all the time because of the dust and noise. It is a wet and dirty job, "mainly making sure nothing goes wrong," as one conveyor man puts it. In the flotation section an operator and a helper look after both the nickel and the copper circuits of the mill. There are also separate circuits for the different grades of ore from Thompson Mine as opposed to Birchtree and Pipe mines. In addition, the crew includes a pump man, who is responsible for the fifty pumps in the building, and two labourers, who constantly "float" during each shift, doing clean-up and relief work.

Besides the crews on shift work, there are a loader-operator who works daily doing clean-up on the grinding and pump floors of the mill and a crane man who operates his crane from a cab or by radio control from the floor, mainly moving heavy equipment, supplies, and parts. Mill maintenance includes carpenters, electricians, mechanics, pipefitters, welders, and their apprentices. Their work is varied. One major task, taking a maintenance crew of five men sixty hours, is replacement of the "lifters" in a rod mill; these are the steel devices inside the mill used to lift the rods as the mill turns to crush the ore.

Like Copper Cliff, Thompson Mill has a strong, musty smell of rock and a good deal of noise, but the general conditions are much better. There is good light and relatively little dust or gas. Because it is comparatively small and better to work in than other surface plants, the mill is the preferred workplace of many Thompson surface workers.

Clarabelle Mill

Only 235 people are employed at Clarabelle Mill: 85 hourly operators with nineteen supervisors and 116 hourly maintenance people with thirteen supervisors, and two senior supervisors. Its capacity is 30,000 tons a day in a continuous process. All operations are monitored in the control room (except for the rail cars) by a foreman and a junior foreman. This modern mill, opened in 1971 at a cost of $80 million, is highly automated, as indicated by the very high ratio of maintenance workers to operators. It is 85 per cent electronic in its operations. The function of the control room is to monitor addition of reagents, cell level control in flotation tanks, and on-stream analysis, all computerized operations. In addition, at the console the foreman controls start-stop buttons for all operations including crushing and grinding, flow and levels of concentrate, and waste disposal. Eight scanning television cameras in the control room monitor all the key stages in the process and also the main gate. The console operator can radio instructions to men on the floor below.

Only a handful of women work in the plant, almost all with low seniority and jobs as "helpers." Although workers can be trained for most of the jobs in the mill in a few days, it takes much longer before they begin to feel comfortable because of the scale of the operation and the lack of specific operating jobs. For many of the operators and maintenance men, the actual work for the day involves only about four hours' labour; the automated system demands little labour but requires their availability.

Ore arrives at the mill by train and is unloaded on a rotary car tipple that dumps the one-hundred-ton cars into a bin. Three men are required for this: a control room operator, an engineer for the train, and a spotter. Two cars are dumped at once, and about three hundred cars are handled a day. The ore is then transported on an automated conveyor system to the 2,500-ton crusher feed bin. The crusher has a gravity flow system of six lines dropping down two levels, one a screen and the other a crusher. The upper level is operated by one man while the crusher is tended by another operator. The crusher is a dirty, noisy place six stories high surrounding the open centre of the crushing plant.

The crushed ore moves next to another plant containing five rod mills and five ball mills where it is mixed with water and reduced to a fine powder. The ground ore is then fed into rows of twenty-two magnetic drum separators to remove iron and twenty-two flotation cells, where reagents are automatically added and the metallurgy of the slurry is monitored. All these operations are automatically regulated from the control room.

Most of the work in the plant consists of maintenance of equipment, mainly checking the rod mills and ball mills. Rods and balls must be changed periodically as the need arises. Some workers do construction-type labour, making adjustments to the plant or equipment. Many of the men appear bored, having little to do but machine tending and monitoring.

None of the milling operations is very labour intensive, but there is a significant difference between the three mills. The character of the operators' work is moving from art to science or, put another way, from a craftsman's control over the process to machine monitoring. With greater automation there is less need for operators, and maintenance takes on a greater role, shifting the operator-maintenance ratio from 1.3:1 at Copper Cliff to 0.8:1 at Clarabelle. Thus fewer workers are needed, but a larger proportion of them are maintenance workers. The newer the plant is, the greater the capital investment and the smaller the need for labour. At the same time management gains more control over the various work processes because it can regulate more closely the actual performance of machinery. Automation seems to increase the isolation of workers; there is less feeling of a working group. Distance develops between individuals when a worker is contacted by radio and watched by a monitor; a closer relationship is possible when instructions are passed on in person. Although automated mills call for less physical effort, they also permit less involvement with the actual process. There is less opportunity for the worker to take pride in the realization that he is affecting the quality of production. The impression one gets from the workers is that they are more closely identified with Copper Cliff as an institution than with Clarabelle.

The Smelters

Inco has two smelters in Canada, one in Thompson and the other near Sudbury. The latter, still much as it was when constructed in 1930, is the largest in the world. On a smaller scale, the Thompson smelter uses essentially the same processes as the original Sudbury smelter. In it the operator-maintenance ratio is substantially lower than it is in Sudbury, 3.4:1 compared with 5.8:1, but this is still much higher than the ratio in the mills, which tends towards 1:1. These figures strongly suggest that smelting is a much more labour-intensive operation than milling. Although smelting has always been highly mechanized, virtually all the major processes remain unautomated, requiring operators to make decisions on the spot and control the machinery. Of all the surface operations of Inco, it is smelting that has been most resistant to capitalization. Or, from another viewpoint, of all the surface operations it is the most likely target for future capitalization. Should processes like those developed for refining (to be discussed shortly) be implemented, well over a thousand Inco jobs in Sudbury could be eliminated.

The material for smelting is the concentrate received from the mills. With the water removed it is roasted and melted at extremely high temperatures. This burns off much of the sulphur (hence the sulphur emissions and tall stacks) and produces a high-nickel molten matte separated from the slag, which becomes refuse. The furnace matte (about 25 per cent nickel) is transferred to hot converters where air is blown through it and waste materials (slag) are separated from the final matte. What is left after about twenty-four hours in the converters is about 75 per cent pure nickel, 20 per cent sulphur, and varying amounts of copper, cobalt, iron, or precious metals, depending on whether the ore is from Sudbury or Thompson.

As in all other Inco operations, workers in the smelters can be categorized by job classifications with varying degrees of supervision and direction exercised. The following are some basic job classifications in the smelters:

Supervised by shift boss and does not exercise direction	Supervised by shift boss and directs others (hourly paid)	Supervised by shift boss and directed by hourly paid workers
crusher operator	thickener operator	thickener operator helper
loader operator	roaster operator	roaster helper
ladle man	labour boss	matte man
crane man	furnace operator	furnace operator helper
bail man	flotation operator	converter man – furnaces
anode mould man	converter operator	converter operator helper

It will be evident that many men in the smelter work in crews around various pieces of equipment. The furnaces, the converters, and the anode casting areas are the nodal points, and the social organization grouped around these pieces of equipment is the unit on surface most similar to the traditional mining crew. There are, of course, major differences, but there are also some important similarities. Whatever the initial impression, the machines do not dictate to the smelter workers as they do in, say, assembly-line production. For example, the workers are required to make many judgements about the progress of the matte. The amount of heat or time are not the only criteria: they are often dealing with concentrate of quite varied types, and they must learn to judge the appropriate colour and cohesiveness of the matte, to decide when to tap their furnaces and dump their ladles. In the parlance of surface workers, there is still a good deal of "art" in smelting, at least for a few skilled smelter men.

Copper Cliff Smelter

A hell-hole, referred to by the *Canadian Mining Journal* (certainly one of the company's friends) as one of the "dark, satanic mills," Copper Cliff Smelter is Inco's largest surface operation and employs about 1,650 hourly rated workers, including 1,380 operators, 240 maintenance men, 27 power supply men, and about 200 staff. The very high ratio of operators to maintenance shows the labour intensity of smelting. This building is the base for the famous 1,250-foot Inco "superstack" (see page 249). The smelter itself has not experienced any radical changes since it opened in 1930, although some adjustments have improved its capacity and efficiency. Heat generation from the gas and Bunker C oil fuels has been increased by the use of oxygen. Efficiency has also been improved because higher-quality concentrate comes to the smelter as a result of the removal of iron ore before it reaches the plant. Matte with a higher nickel content can be made from fewer tons of ore.

* * * * *

A Sudbury smelter man with twenty-two years' experience:

It's so cold your feet are tingling. It must be twenty to thirty degrees below in winter. . . . Everything you touch is just frozen; it's really unbearable. Sometimes you wonder if it's worse in the winter or worse in the summer. It goes from one extreme to the other. In the winter you get colds.

You are in the heat one minute, the cold the next. And if you get rid of a cold, one day at work and, bang, it's back. . . . You have to have a certain amount of draft in there to send the gas out all the time, winter and summer. If you close that building in you're going to suffocate.

* * * * *

Over all, about 80 per cent of the current jobs are essentially the same as they were in 1930. Some of the menial bullwork has been eliminated, but a great deal persists. For example, twenty men once worked at keeping the area clean near the furnace, but this job is now done by a front-end-loader crew in half an hour once a week. Power floor sweepers have also cut down on in-plant clean-up, at the same time decreasing the dust. The job of twenty-four "crane chasers" who place crane fasteners on the ladles of molten matte is scheduled to be eliminated by equipment capable of doing this job. On-stream analysis, like that at Copper Cliff Mill, is scheduled for the sampling and will record constant results in a control room. This, of course, will also reduce the labour needed for sampling.

Improved tools, such as the impact wrenches now used by maintenance workers, reduce the amount of labour needed and increase efficiency. Another innovation is the tugger hoist, an air-operated machine used to move material to the top of the smelter, a job previously done manually with a set of pulleys. But none of these labour-saving devices has fundamentally changed the nature of the work.

At the smelter material is received from the mills and conveyed to the top of the building, where it is fed into bins for the thirty-eight multi-hearth Herreschof roasters (twenty-one feet in diameter and thirty-eight feet high). Each roaster has a drying hearth and ten interior hearths with "rabble" arms continually stirring the concentrate and guiding the material to the hearth below. The top floor is very hot, but as one moves down closer to the bottom of the roasters the temperatures increase to the point of being unbearable. A feed tender works at the highest level, ensuring that an adequate amount of concentrate is deposited in each roaster. On the next five floors are the furnace men and their helpers who operate each battery. Their task is to tend and regulate the heat of the furnaces. It is *very* hot on each of these floors.

The roasted concentrate is delivered to the reverberatory furnace (114 feet long and 30 feet wide). Here slag is removed by the slag skimmer, whose job it is to pour the slag from the furnace through a tap hole into waiting slag cars. Each of the twenty-eight twin-pot slag cars holds thirty-five tons, and when they are dumped onto the slag heaps an orange glow familiar in Sudbury occurs. Another worker known as the slag keeper regulates the amount of slag let out of the furnace and closes the hole with a ball of clay about eight inches in diameter (see Figure 19). He reopens the hole when the slag pots have been changed. He makes his own clay "stoppers," put in place with a long metal pole. Although his job working alongside the furnace makes his face hot, behind him is an open area that makes his back very cold in the winter.

A furnace operator removes the molten matte (temperature 1,150°C.) from the reverberatory furnace into ladles, which are transported to the converter aisle by six cranes. A bail man (or hooker, as the smelter men call him) works on the floor between the furnaces and the converter aisle attaching the hooks from the cranes to the ladles.

* * * * *

A Sudbury bail man:

It's an awful dangerous job down there. One day the crane man picked up this ladle, and as he brought it over it started coming down. I was waiting for it. The crane man says, "Get out!" The way the ladle was, it would hit the floor so it was going to tip, so I ran into the tunnel. Sure as heck it did hit, but it didn't tip. It just hit enough that it splashed all over in the area, maybe ten or twelve feet around, but stayed there. The crane man didn't know what to do. He thought I was caught. He called out, and they looked for me. They couldn't find me. I was way down at the end of the tunnel. They looked around with a flashlight after calling, but I was shaking so hard I couldn't answer.

* * * * *

Fig. 19. A tapper's helper on the job

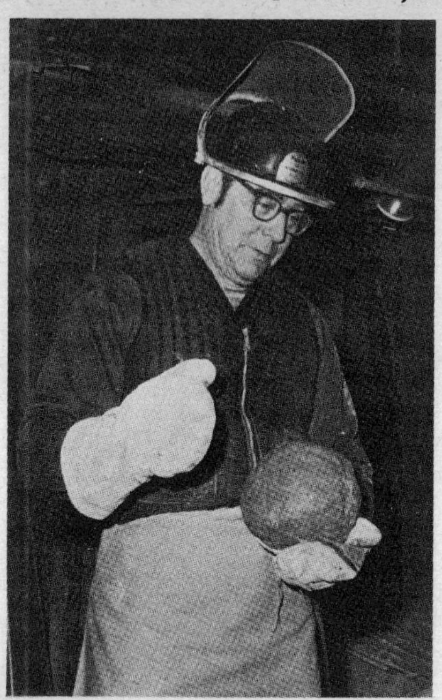

Inco Triangle (March 1977)

The converter aisle is 1,450 feet long and lined with nineteen converters. Seventeen are currently in operation, five on copper and twelve on nickel. Each is thirty-five feet long and thirteen feet wide, holding one hundred tons (every ladleful makes four inches). The job of the skimmer is to remove the slag and pour out the matte, judging by the colour when separation has taken place (see Figure 20). He heats the material to 2,200°C. and turns the heat down, letting it sit for about ten minutes, and then begins to skim by moving levers that tip the converter. He decides how much furnace matte to add or slag and finished matte to remove. Training for the job takes eight weeks, but most skimmers have had much longer experience with the equipment before taking the controls. Skimmers have a top-paying job, and it is one of the few surface jobs on bonus (ranging from 16 to 27 per cent or a value of $200 to $300 a month). The skimming bonus is based on the number of tons of matte produced and recorded for all the skimmers working on each of three shifts. This is management's recognition that skimmers, like miners, can control the rate of production and serves as an inducement to maximize production.

* * * * *

A Sudbury skimmer:
I didn't even want skimming. I never thought I would ever get to be a skimmer. The boss said, "Well, you are the next one up in line for skimming." So I says, "No way." He said, "From now on you are going to break in skimming." I didn't want to say no so I broke in, and they put me on. I didn't want to let my boss down.

Is skimming considered to be one of the most prestigious jobs in the plant?

Oh, I think it is. It really is. We're under strain all the time because you don't know sometimes what you have got in that converter. Your boss says, "Well, why don't you take a pot of this or a pot of that," and you don't know what reaction you are going to get when they put it in. Anything can happen. I've seen converters blow up and just throw the whole batch right out. Not long ago they were putting

Fig. 20. Pouring molten slag from a converter

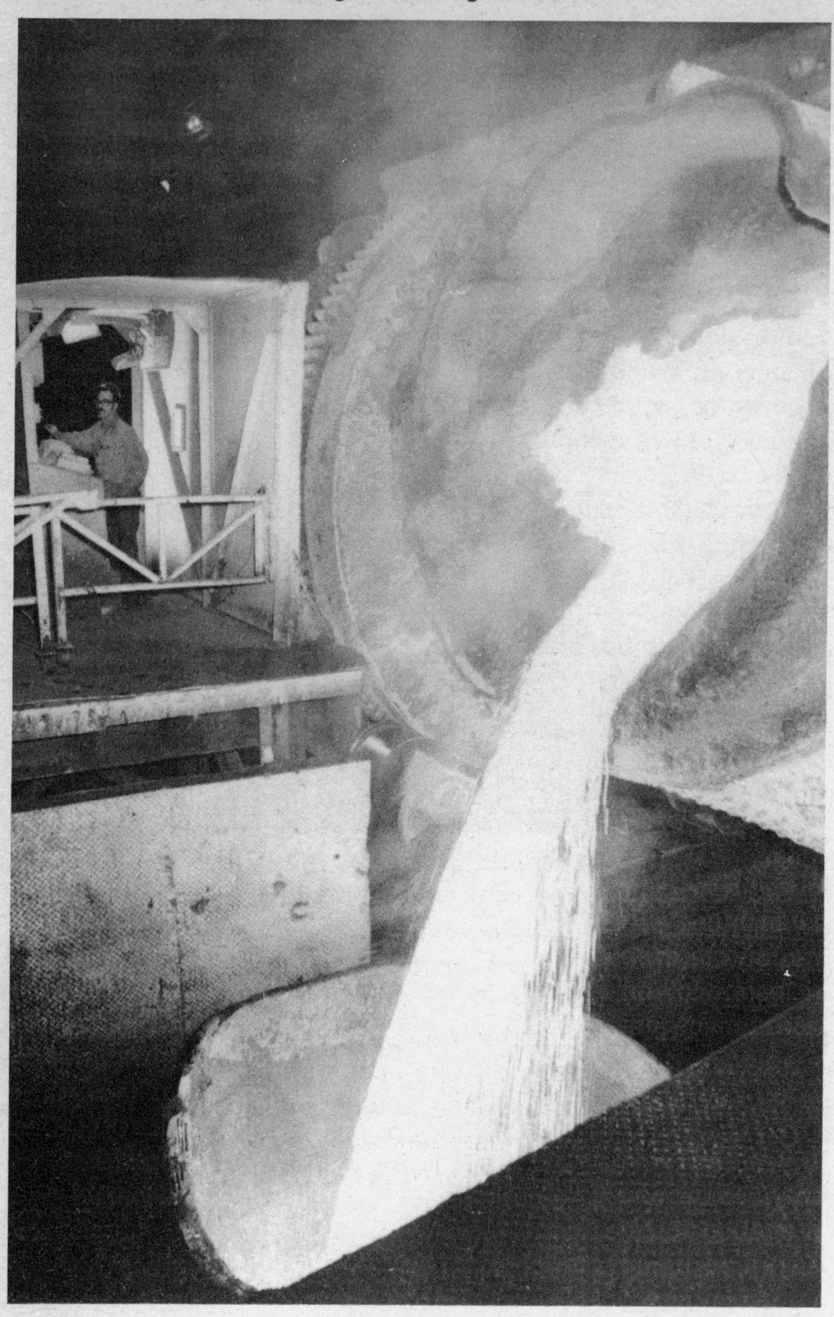

IN Manitoba (October 1978)

coal matte in those barrels into the converters. They put them in and half one end blew off, all the brick and everything inside. It darn near killed a man. We never know what we are going to get. Shop cleanings, they call them; stuff from all over the place. I don't know what it is; nobody does. Just put it in the pot.

* * * * *

Skimmers work immediately alongside the open molten metal and have the hottest jobs in the smelter besides being exposed to a great deal of gas. The molten matte is at 2,200°C. when handled by the skimmer. Behind the converters, working along with the skimmers, are the punchers, ensuring that air flows into the converters at a rate of about 20,000 cubic feet per minute. The air is required to oxidize the iron and sulphur in the matte.* At one time these men used bars (one holding a bar and another hitting it with a hammer) to keep the air holes on the back of the converters open. Later, pneumatic hand drills, and after 1968 automatic pneumatic punchers (known as tuyeres), were introduced, reducing the number of men from four to one on each of the nickel converters. These machines work only up to a certain point, however, and at the white-metal stage it is necessary to revert to hand punching. Four punchers work on each of the five copper converters because the automatic punchers are not able to prevent the plugging. When the copper sticks to the bars of the punching machine during the last four hours of converting, they are removed and the work is done by hand with a hammer and bar. As a former puncher says, "The punchers are just beat taking that heat. Oh, it's awful. . . . When you are punching on the nickel machines it's not so bad, but in the copper end they have got to slug away for four hours straight, pounding with bar and hammer. It's a four-pound hammer, and then they've got to drill all those tuyeres out every time they make copper." As might be expected, punching is considered the lowest job in the smelter, and there is a high turnover.

From the converter, the matte is poured into refractory moulds of which there are 271, each holding twenty-five tons

* This is the process developed by Henry Bessemer (1813–98) that revolutionized the converting of metals.

of matte. These are twelve feet long, eight feet wide, and two feet deep. They are moved by an overhead crane. Once cooled to 400°C., the matte ingots are broken up first by a ten-ton pile driver, then by a 48-by-66-inch jaw crusher, and finally by a series of crushers, after which the matte is passed to the matte processing plant.

This plant is part of the Copper Cliff Smelter complex and continues the matte separation. It receives the matte and crushes it further in rod and ball mills (much smaller versions of those found in Clarabelle). Some of the metallic material is removed by a magnetic separator and then goes into basket centrifuges. The centrifuges are operated by men who watch the spinning material until it is dry (usually seven to fifteen minutes, by the operator's judgement). Two men work alongside one another at separate centrifuges. While one machine is drying, the other is being filled. The operators are able to spell each other for lunch and breaks, but the noise precludes much communication. These men can be trained in one or two days. Although their working conditions are noisy, they do have some control over the pace of their work. The dry metallics proceed to the Copper Cliff Nickel Refinery.

The other materials continue to be ground and subjected to froth flotation to separate the copper and nickel sulphides (the copper sulphide being transported to the Copper Cliff Copper Refinery). The flotation process is overseen by a flotation operator who has two helpers. He decides, on the basis of assay sheets prepared every two hours and the colour of the froth, how much reagent and lime to add to the tanks. His job requires a good deal of knowledge and experience, and learning it takes a minimum of six months. The matte separation operation is relatively labour intensive, employing about 60 per cent of the 314 people working in the matte processing plant. The flotation process is continuous for the most part (the centrifuge is more of a batch process). The sampling and analysis are all carried out manually.

In contrast, the two other areas of the plant are automated; about 114 people work in the fluid bed roasting process and normally 32 are in the chlorination-reduction area. About eighty of the employees in the fluid bed roasting area are actually shippers, sending some of their Sinter 75 nickel to market and the rest to Port Colborne and the United Kingdom for refining.

Both the fluid bed roasting and the chlorination-reduction areas have their own control rooms. A control-room operator requires about six months' training before becoming an independent worker (although a practical knowledge of the plant is essential first). The roasters in the fluid bed roasting area generate a great deal of heat. There are four of these, two re-roasters, and four coolers. Water is added during the process to produce pellets. The chlorination-reactor makes high-purity nickel oxide by treating the sinter with gaseous chlorine in 1,200°C. fluid bed roasters. These are other end-products of the Copper Cliff Smelter (see Appendix IX).

Thompson Smelter

The Thompson Smelter has gone farther than Copper Cliff in introducing innovations that reduce the amount of labour in its operations, but, like all smelters, this one is characterized by heat and the handling of molten metal. About 340 hourly production employees and another 100 maintenance men are employed. Nickel concentrate is pumped in slurry form from the mill to eight thickener trays, some eighty feet in diameter, where the water content is reduced from 75 per cent to 30 per cent; it is further reduced to 13 per cent in drum filters. This area is highly mechanized and requires only one thickener man and one pump man a shift. The dried nickel concentrate is then placed along with some sand flux in rotary bins, which feed the five fluid bed roasters. This is the only automatic part of the process. The dryers are endo-fueled (providing their own fuel), burning up about half the sulphur in the concentrate at temperatures from 1,100°F to 1,275°F. This process is semi-automatic. It is operated from a control room where the temperature and feed rates are set, but the roaster and furnace operators tell the control room the temperature and amount of power they want. (As a former control-room operator said, "It's a boring job; I went stir-crazy.") The smelter gases pass through electrostatic precipitators, where dust is removed, and finally up a five-hundred-foot stack.

The material is then melted in one of five electric furnaces, after which it is known as calcine. Each furnace, tended by five men, is three stories high. Here the molten matte is formed and the waste rock becomes slag. Heat is generated by an electric current flowing between six carbon electrodes, each fifty feet long and twenty tons in weight. Temperatures of 2,300°F. are produced. This operation (along with the refinery) uses enormous quantities of electricity. It consumes about half as much as is required for the entire city of Winnipeg (some 240 megawatts). Electricity is bought from Manitoba Hydro at very economical rates negotiated when Thompson was built. These rates were scheduled to last until 1981, at which time they are expected to triple (unless, of course, the Lyon government of Manitoba can be convinced otherwise, which it appears may happen; see page 354).

The furnace area in the plant is hot but not as bad as in the Copper Cliff smelter; nevertheless, it was necessary to replace glass windows with louvres because workers kept breaking them (at a rate of about six thousand a year) to let in cool air. The furnace is an enormous operation, with five men working at each. The most senior is the furnace operator, who watches the furnace and checks the matte and slag levels with a bar, making the decision when and how much to tap. Skimming is not done until at least forty inches of matte have formed on the top, and tapping needs at least thirty inches of slag on the bottom, otherwise the valued matte may be thrown out with the slag. The furnace operator is responsible for overseeing the matte man, the pump man, the skimmer tapping slag at the back of the furnace, and a helper at the front who helps the furnace operator tap the matte. The Thompson skimmer, after making an opening in the back of the furnace with an oxygen torch (unlike the Sudbury operator who uses plugs and pours into train pots), skims the molten slag down a chute to a granulator which sprays the slag with water and then pumps the slurry out to the slag pile. He must protect the chute with a paste lining and rebuild it every shift. The furnace skimmer watches gauges for the water pumps, furnace levels, heat, and slag-line pressure and takes a sample of each skim. Skimming time varies. Sometimes it goes on for the whole shift if the slag level is high, at other times for only about four hours.

At the front of the furnace the matte is tapped into ladles for the converters. On the smaller furnaces the furnace operator uses traditional clay plugs to close the tap on the front manually, but on the larger furnaces plugs are inserted mechanically using compressed air punchers (because the force behind the matte would require about twelve men to make the plug hold). On the converter aisle, skimmers for each converter use the traditional Bessemer process of blowing air through the molten matte. There are also three crane operators working eighty feet above the aisle. Each crane has a sixty-ton capacity and lifts the twelve-ton ladles with up to eighteen tons of molten matte. Bail men direct the crane operators with hand signals.

When converting is complete, the molten Bessemer matte is poured into ladles for transportation to two 110-ton holding shells. The matte is kept hot by oil burners under the shells until it is time to cast the anodes. The process here is to pour matte from the holding tanks into a mobile ladle car that holds four tons of matte. The car pours the matte into forty-seven anode moulds by moving along a railway track beside the moulds. Each is filled to a depth of 2 1/4 inches after being coated with a silicate substance to keep the matte from sticking. The matte solidifies into 560-pound anodes that are picked up by a gantry crane and cooled for twenty-four hours in annealing boxes, at which time they are ready for the refinery. These anodes are made round the clock (460 each shift). The workers in this process include the ladle operator, who sits on the self-propelled ladle car; a mould attendant, who smooths the surface of the anode after it is poured; a hooker, who helps lift the anodes; a fork-lift man, who moves them away; a hoist operator; and a wash man, who puts silicate into each mould before the anode is cast. This is a steady operation where the men work as a team. Each has his specific task as they co-ordinate their work to produce the anodes destined for the nearby refinery.

Contrasts between the Copper Cliff and Thompson smelters are really not very great. The major differences occur in the final stages, after a nickel matte has been produced in the converters. At Copper Cliff the matte goes into a fairly automated process that produces some final products as well as feed for the refineries. At Thompson all the matte is made into moulds for the refinery, and the process is basically manual with machine assistance. General conditions in the two plants differ somewhat because the older Copper Cliff operation suffers more from heat, dust, and gas than the one in Thompson, but the actual processes differ little. The heat in Thompson is mainly electrically produced, and the furnaces are better insulated to reduce the amount of heat in the work areas. Operations around the furnace are also somewhat more modern in Thompson, such as the continuous slag-skimming process that granulates the slag before moving it to the slag heaps, in contrast to the pots that carry molten slag by train to the slag heaps at Copper Cliff. The Thompson operators have a somewhat more mechanized method of closing the taps at the front of the furnaces, but generally the structure of work is the same. There is also very little to differentiate the converter operations in the two locations, both relying on skimmers and punchers.

There was very little advance in the technology of smelting in the thirty years between the building of the two plants. Work remains labour intensive, and working conditions are difficult in both. The organization of work for smelter men has changed very little over the past forty years. Developments in refining represent a stark contrast.

The Refineries

Inco has three nickel refineries in Canada. The difference between the Port Colborne Nickel Refinery (PCNR) and the Thompson Refinery is similar to that between the Copper Cliff and Thompson smelters: there were some technical advances in the forty-three years between their construction, but they were relatively minor. Both use an electrolytic refining process, and both are labour-intensive batch operations. The newest refinery, the Copper Cliff Nickel Refinery (CCNR), represents a radical restructuring of the process and organization of work. It uses a high-pressure carbonyl process that is highly automated, requiring very little labour. The difference here is even greater than that between the Clarabelle and Copper Cliff mills.

In addition to the three nickel refineries, Inco also has one copper refinery, the Copper Cliff Copper Refinery (CCCR). Like the Port Colborne and Thompson nickel refineries, CCCR uses an electrolytic process and has many similarities with these plants. In 1973 an "electrowinning" process was added to CCCR. This was an attempt, much as CCNR was for nickel, to restructure the processing of copper. It involves not only new processes but, like the CCNR, a new form of training or "people technology" as Inco officials like to call it. As will be seen in Chapter 8, the modular training system originally adopted in the electrowinning section of CCCR and then applied on a large scale at CCNR has now been adopted for the entire Ontario division of Inco.

The three electrolytic refineries have many job classifications in common. As in the other operations, these are grouped into three types by their relationship to supervision and direction:

Supervised by shift boss and does not exercise direction	Supervised by shift boss and directs others (hourly paid)	Supervised by shift boss and directed by hourly paid workers
tube filter man	weigher	utility man
soda ash tank man	stripper	tank repairman
liquor chaser	section leader	stripper helper
front-end loader operator	purification man	shear floor packer
kiln operator	press man	plating tank man
filter operator	precipitation man	lift truck operator
employee equipment man	cobalt treatment man	labourer
crane man	cathode shear operator	filter press cleaner
cobalt calciner man	carbonate man	box man helper
matte production man	box man	anodes scrap man

The relationship between the workers is somewhat different at CCNR and will be discussed in some detail in that section. The four refineries provide an opportunity to examine a number of important developments in the surface operations. Since 1973 two new processes have been added, each with differing results, and what was once a major plant has been moth-balled. It will be possible to understand some of the reasons behind these major management decisions and their implications for the nature of work on surface.

All four refineries have the same objective: the production of high-quality nickel or copper for the market from matte produced in the smelters. Appendix IX lists their products. The refining process consists of the separation of various metals from the matte and, by either electrolytic or carbonyl processes, reduction of the impurities in the product. It is essentially an up-grading process which also allows the metals to be isolated in pure form.

Port Colborne Nickel Refinery

The major product of the Port Colborne operations has been Class I electrolytic nickel (over 99 per cent pure) since it was first introduced in 1926. Its raw material (or feed) is nickel oxide, a fine powder received from the Copper Cliff Smelter and shipped by rail. The plant itself is much as it was when constructed in 1918. The equipment is original even though there has been continual rebuilding. The layout of the buildings and flow of material is old; even the structure shows signs of wear. Wear is an important consideration in the electrolytic process because of the high acid content in the vapours given off and their effects on the building's steel and concrete.

As in other older plants, many minor changes have reduced the amount of bull-work, but the actual operations remain essentially unchanged. Two of the most dangerous areas of the refinery have, however, been eliminated. The sintering plant was phased out in 1958 and the calcine department in 1972. Both were major sources of nasal and lung cancer. One Port Colborne employee who worked in the sintering plant said, "You were working three feet from a man and couldn't see him because of the nickel dust." Conditions here were much like those in the sintering plant in Sudbury, closed in 1952. Other less noticeable changes include increasing the size of the refinery's furnaces about fourfold, upping their effectiveness because no more operators were required for greater production, and the introduction of a casting wheel with twenty-eight moulds that increased the pouring efficiency of the furnaces.

In contrast particularly to CCNR but also to the Thompson Refinery, PCNR was very labour intensive. This is indicated by the size of the maintenance staff, which was about a third the size of the operating staff.* Many aspects of the operation could be further mechanized or automated, but, as the Port Colborne management stresses, the work cannot be justified by the plant's (and company's) cash flow – or, in other words, management's plans for PCNR.

* * * * *

An Inco official in Port Colborne:
Our process here is very labour intensive. There's not much automation. Mostly all the thing is manual as far as the handling of the product. There are mechanical aids, but people are doing these things as opposed to someone sitting in a control room pushing a button, making things happen. . . . The process is a continuous flow. We have operators who continuously monitor the things, but when it comes to the actual production, the removing of the finished sheets, the transporting of them, the cutting of them, the packing, the loading or shipping, it's all manual.

* * * * *

In the actual electrolytic process at PCNR, nickel oxide is mixed with petroleum coke and heated to 2,800°F. using Bunker C oil, so that it liquefies and is cast into five-hundred-pound metal anodes. In the summer of 1977 only two furnaces were operating in the foundry and three were shut down. It takes about ten hours to tap a furnace, and about twenty men are involved. Three are directly engaged in tapping the furnaces to allow the liquid to flow. Working alongside the furnace are a furnace man and his helper, a gas man, and a tap-hole man. In an enclosure is the wheel man, who is responsible for controlling the movement of the casting wheel with its twenty-eight moulds. There are also a pin puller who removes the anodes after the cast has cooled by pushing the anode up with a pin set in the mould, and two men in the bar pit, putting suspension bars on the anodes. Two hoist men remove the

* In trying to understand the job structure at Port Colborne it is often difficult to distinguish the factors that result from the nature of the PCNR operation from those attributable to its being moth-balled as a significant producer of nickel. They come together in a number of ways. The research for this section was done in the summer of 1977, before PCNR was closed.

anodes, two men operate the gas locomotive, and another worker is a track man. The men working in this area are exposed to a great deal of heat and do heavy work.

In the same plant is located the "F nickel shot" process, in which molten nickel is poured into high-pressure water jets to form pellets. This process uses Sinter 75. The shot falls into holding pits and is then dried and packaged.

In the tank house there are 1,640 tanks, each of which can hold thirty-one of the anodes produced in the foundry operation (see Figure 21). Not all the tanks are in use. The anodes are separated by cathode sheets of stainless steel or a thin nickel starting sheet. When a nickel sulphate-chloride electrolyte is circulated in the tank, nickel from the anodes forms on the sheets. The sheets remain in the tank until a sufficient amount of nickel was "grown," that is, has been transferred from the anodes to the cathodes. Every twenty-eight days the anodes are removed (reduced from about 500 to about 150 pounds) and recycled back to the furnaces to be recast after having been through the scrap wash spray pit. In the tank house the anodes are handled by a hoist. Two men check for possible electrical shorts in each section of twenty-nine tanks. Another two work as tank cleaners. The tank house has very high humidity, but the venting and lighting there are all recently installed. Because of the acid vaporizing, the steel structure of the building is slowly eroding.

The major final product of PCNR is electrolytic nickel that has grown on the starter sheets (cathodes) in the tanks (see Figure 21). When complete they have grown to about 135 pounds and a 28-by-36-inch size. The nickel cathodes are sheared into pieces as small as one inch square, depending upon customer specifications. The shearing is a very noisy job (even though sound-proofing has been added to about half the machines). Two men work on each shearer and others in the crew are sorters, weighers, and drum headers.

The printing room begins the process for producing "rounds," or one-inch diameter nickel buttons. Stainless-steel sheets are painted with a dielectric ink, leaving exposed 1,228 circles where nickel will grow when the sheet is put into the electrolytic tank with anodes. In the printing room one man operates the printer and two others clean the sheets with acid. Round-stripping occupies five men who remove the grown

Fig. 21. Electrolytic cathodes and tanks

Inco Annual Report, 1976.

nickel. One man stands in a hammer room with an automatic pneumatic hammer that strikes the sheets and, with the help of another manual hammer, removes the rounds from the sheets. The rounds fall through the floor grille, and often he must use an axe to break them up so that they will fall through. This is a very noisy job.

The rounds are then moved by conveyor to the polishing area. Here there are four men sorting the polished nickel, a job usually assigned to workers with back problems. This is a two-shift operation. There are also a packer, an operator, a weigher, and two men putting the heads on drums.

The foundry additives plant (on the location of the former sintering plant and calcine department) produces batches of nickel containing additives to the customer's specification. It melts finished cathodes and adds magnesium and silicon to the nickel to form alloys, using advanced casting and material-handling techniques. This plant began operation in 1972 at a cost of $3.6 million and will continue even with the most recent layoffs.

The hottest, dirtiest job is in the anode scrap wash spray pit, where valuable slimes are removed from the spent anodes. Here two men work, one using a high-pressure hose and the other turning the anodes. There are many accidents resulting in back injuries here because of the weight of the anodes (150 to 200 pounds), and the anodes, which become brittle when spent, sometimes break and can injure the workers. There are fumes and mists from the high-pressure hoses and mists from the slimes. One Inco official admitted, "These things have caused a hell of a lot of back strains and back injuries because this is a manual operation – they walk the things around; they hose them to wash the slimes down. We have developed a system to mechanically clean the things." This job could easily be mechanized, and it has been considered; but because of the layoff situation there is little pressure to make the move.

* * * * *

A Port Colborne union official, discussing workers in the scrap wash pit:

The workers in anode scrap wash are dressed in rubber outfits and have got to wear a face shield. That place is another source of injuries within the plant. But the way we look at it, god damn it, sooner or later we know there's going to be changes made there. As a matter of fact, I think they're just about to be introduced now, which will eliminate it. Now whether the company utilizes the number of people involved in doing that operation we have no idea.

If you start complaining about the work conditions there may be a technological change.

That's right. There's going to be a method where these people are not doing that kind of work there any more.

So there's a danger on your part if you press for it; the company may eliminate those jobs.

That's right. You try to eliminate one evil and you create another. It's a heavy job, and statistics show that a lot of back injuries come out of that particular job. The floor is wet. People are trying to manipulate heavy anodes by shoving them one way or another, moving them around and lifting them on these flat cars.

* * * * *

Many of the jobs operate under unofficial quotas. For example, workers known as the box gang, who make canvas-sided cathode boxes to enclose the nickel starter sheets, work two to a table (three tables) and must make 110 each day (they work days only). They construct the wooden frames and secure heavy cloth to them. Strippers, who remove thin nickel starter sheets from the stainless steel plates, must also do 110 per shift. Normally two or three men do this job. Then the strips must have straps, used to suspend the sheets in the tanks. Five men operate the strapping machine while another two spot-weld the straps. These men feed the machine that rolls the nickel starter sheets, and they are on a quota of five racks of 240 sheets a day. They are not on bonus (nor is anyone else at PCNR).

* * * * *

An Inco official:
We have a sort of quota system. For instance, on the stripping, a lot of men just couldn't do the job or wouldn't

do it, so the foreman finally said, "Look, unless a man can strip 110 stainless steel sheets a shift, I just haven't any use for him." So anybody beyond that figure kept his job. Well, you can be sure that it wasn't long until this minimum became a maximum. Now everybody does their 110.

Another company official:
We say that we have no work systems here that would involve production quotas. But we do have production quotas. The guys actually cutting nickel are expected to cut so many tons, and they go like hell; but when they approach that quota – unofficial quota – they start to take a smoke or coffee break.

It's more like an understanding than an agreement?

We say there is no quota system here. You come in and we expect you to work for eight hours. Whatever the foreman assigns to you, you have to do. But there is sort of an under-the-table understanding in certain work groups. You make so much or do so much and it is considered a good eight hours' work, even though you may get it done in six.

A union official, asked about the "unofficial" quota system:
The company will never agree with you if you say, "Well, you got your people on the quota system." They say, "Oh, no!" But if an individual doesn't get out X amount of work, he can be reprimanded; if he doesn't build 110 boxes. The tank cleaners have to clean so many tanks. The unit men have to look after so many units. Everything is numbered. The shearing department, they have to cut X number of thousands of pounds of nickel per shift. I think now it is up around 108,000 or 110,000 per shift. And it's competition that the company creates on each shift. For example, your shift may cut 106,000. My foreman comes in following you and he says, "Well, they cut 106,000; we have got to try and get 107,000." You see, that is the kind of competition that's generated within the plant.

* * * * *

As mentioned earlier, since September, 1977, the core of the Port Colborne operation has been closed down, and there is no indication that the plant will do anything but produce a few specialty nickel products (such as rounds and nickel with additives) and act as a storage depot. The impact of the layoff will be discussed later, but the reasons for it are already evident. Not since 1971 has PCNR hired anyone (aside from eight or nine apprentices), and in 1972 about three hundred men were laid off, so that the massive 1977 layoff affected even those with very high seniority. Instead of following the policy at the other older operations, such as Copper Cliff Mill and CCCR, Inco has made little attempt to rebuild and maintain the PCNR operation. It is no coincidence that CCNR came into operation in the same period. As one Port Colborne official said, "They produce something between 1,400 and 1,800 pounds per man-shift and we are fighting like hell to get 400." Moreover, CCNR's costs are primarily in capital equipment with little variable cost for labour and the materials, such as the reagents, that are consumed in the electrolytic process. Inco officials have decided in their program of capitalization to put more weight on operations with high capital costs than on those with high labour costs. This has doomed the Port Colborne operation, and with it about two thousand jobs.

Thompson Refinery

The refinery in Thompson uses an electrolytic process like that at PCNR. It is also labour intensive, but many of the subprocesses have been either highly mechanized or contracted out to private companies, reducing the overall size of its labour force in comparison wih PCNR. About 230 hourly rated production workers are employed in the electrolytic refinery, and much of the maintenance work is done by a separate department, as is the sampling.

The process at Thompson begins when a crew of two strippers, a shearer, two loopers, and a loop maker makes starter sheets by removing 1/32-inch nickel sheets from stainless steel blanks that have been plated for twenty-four hours. These starter sheets, which are simply thin nickel sheets with metal loops on top to hold them, become the cathodes suspended in canvas-covered wooden boxes in the tanks where the final product, nickel sheets, is produced (see Figure 21). This is the same electrolytic process developed early in this century and also used at PCNR. Each plating tank has twenty-nine anodes and twenty-eight cathodes, the anodes coming from the casting area of the Thompson Smelter. When the starter sheets are in the boxes, the electrolyte flows through after dissolving the nickel from the anodes and deposits it on the sheets. Sulphur and precious metals remain at the bottom of the tanks as sludge and are periodically removed by tank-cleaning crews.

Cathodes are left in place for seven days, growing to about 140 pounds and a thickness of about half an inch; the 550-pound anodes are used for seventeen days, being reduced to about 200 pounds. The cathodes become a finished product, and the anodes are recycled. An electrical current of about 10,000 amperes flows through the electrolyte. The canvas-covered wooden boxes become worn out from use, and nickel contained in them is recovered from the scrap by burning it in a dump, collecting the ash, and returning it for smelting. This procedure recovered 250,000 pounds of nickel in 1977 and 500,000 in 1976. (It also helps explain how some unusual materials turn up in the converters.)

Thompson, unlike PCNR, contracts its box-making to a local firm and effects a substantial labour saving within the plant. Other differences include hydraulic openers on the presses used to squeeze foul electrolyte out of used boxes, requiring less work than those at PCNR. Greater efficiency in the use of precipitate (the reagent used to separate nickel solids from the electrolytic solution) at Thompson has also resulted in decreasing the need to clean presses from fifty to only six a day. Cleaning presses is a very dirty, heavy job, and many people who do it get nickel rash, a skin infection caused by the hot steam used in the area and the nickel from the presses. Not only have all these changes lowered substantially the labour and maintenance costs but they have also increased the plant's capacity by 100 per cent, mostly because of filterable precipitate.

* * * * *

A Thompson refinery worker:
They used to have a whole gang of men doing the boxes in the plant here, and then it came about that they phased

them out. But when it was done, nobody was really concerned because there were so many people coming and going. There was no hint of a layoff or anything like that. Workers were just reassigned to other jobs. There was no problem. No big thing. In fact, you find very, very few complaints about technological change.

* * * * *

When the finished nickel cathodes are removed from the boxes, they are moved on a narrow-gauge railway to the shipping area where they are bathed in acid and cleaned. They are then bundled into 4,400-pound piles, strapped, and moved off by fork-lift trucks. Unlike PCNR, Thompson does little shearing of cathodes. There are two shearing machines, but most of the nickel is shipped as 28-by-40-inch cathodes.

During the refining process several by-products are produced. At Thompson, hydrogen sulphide, a highly toxic gas, is used to precipitate copper and sulphides, which are filtered out of the electrolyte and sent on to Copper Cliff for further processing. The section of the plant where this is added is restricted, and only workers wearing masks can enter. The process is automatic and operated from a control room. Also removed from the electrolyte by the addition of chlorine are cobalt and iron. The cobalt is finished here as a cobalt oxide and sent to market.

* * * * *

A Thompson cobalt treatment operator:
It's batches, but it is also continual in the sense that you have to keep taking things out. You just take a batch and inject gas into it up to a predetermined level so certain chemical processes take place. You have meters and measure it and add other acids and things and finally it is ready. Then you filter it out, and that separates the iron; and then you add other reagents and filter it out, and that removes the copper, and so on. It can take more than a shift. There's a pretty high time variable there, really. You can do a batch as quickly as four or five hours, but on the other hand it could easily take you ten or twelve hours. It

depends on how concentrated the solution is and a lot of things.

Is it an interesting job?

I find it interesting because I make a game out of it; a lot of people complain because it's a pretty common job. But my job is more interesting than most because it's easier to make little innovations. It is a batch, and you are in full control of it. You are not responsible to a control room. There is no control room or mechanism monitoring what you are doing, and you are left alone. So, you have a lot more opportunity to play little games with yourself and with the batch. And you also have the opportunity that if you are not feeling well, you don't have to work so hard; on other days, if you feel a little energetic, you work more.

* * * * *

Another area of the refinery is scrap wash. After the 550-pound anodes have been in the tanks for seventeen days they are pulled out, then weighing about 200 pounds. These are recycled by the scrap wash crew. The spent anodes are delivered on racks to their area. One crew member pulls the covering bags off the anodes, another lifts them onto a tipple table with the aid of a hoist, and a third breaks them with a sledge hammer. The scrap then falls onto a long conveyor belt that returns the material to a crusher from which it is eventually fed back into the smelter. Another person in the area repairs the bags. The scrap wash crew rotates its tasks.

* * * * *

A Thompson scrap wash worker*:*
We rotate on the table. Say you start off pulling bags; throughout the day it will end up that three people on the table will each take a turn on bags. It's easier, because for somebody to break with the sledge all day, if it's running heavy, there's just no way. . . . Let's say he's broken two hundred anodes in a two-hour period. . . . If you start off pulling bags in the morning, you pull until the guy on the sledge hammer has broken his share of anodes and then you rotate to hoist man. The hoist man goes to the hammer. The guy on the hammer comes over and pulls

bags, etc. . . . It gives everybody a break and it breaks the monotony of the job, because it's a monotonous, boring job.

* * * * *

This is one of the few jobs involving heavy work that women have been permitted to do in Thompson. The following is one woman's account of her introduction to the crew:

There's basically no real training in it because you learn it on the table. I was so terrified at that table I think it took me two months just to relax enough to learn how to pull the bags properly. I used to come home and I was so knotted up in my back I used to have to take pain pills because it was just driving me crazy. It was nerves. At that time, putting women on the table, men didn't particularly care for it. . . . I stayed on the hoist all day, and it never bothered me; and then the men would just go back and forth between the hammer and pulling bags, which I felt evened it out a bit from my side. I didn't feel such a pain in the neck. After a while, one of the guys says, "Well, why don't you try the hammer?"

I said, "I've gone this far, why not?"

* * * * *

Generally, the differences between the Port Colborne and Thompson electrolytic refineries can be attributed to their relative positions within management's planning of the company. The Thompson Refinery is integral to an overall complex; PCNR has become redundant. Many of the advances made at Thompson could as readily have been adopted at Port Colborne, increasing its efficiency and eliminating many undesirable jobs, but the company did not plan them because it did not want to add to its capital expenditure. The workers did not press for adoption because it would have meant the immediate loss of some jobs. In Thompson, on the other hand, the turnover is so high that the eliminated jobs are hardly missed, and workers are simply re-assigned. For example, the union there did not even resist the contracting out of box making, which at Port Colborne would have meant the direct loss of jobs.

Copper Cliff Copper Refinery

Built in 1929 by the Ontario Refining Company, Copper Cliff Copper Refinery has retained essentially the same character since that time, although there have been some important developments. Output of the plant has steadily increased. Its 1930 capacity was 300 million pounds of copper annually, and during the war it was expanded to 325 million pounds; in 1970 it was further extended to about 400 million pounds. The tons per man-shift increased steadily from 0.34 in 1931–35 to a peak of 0.51 in 1956–60 but have since declined to 0.47 for 1971–74. The recent decline reflects, in part, problems with the new electrowinning process. Of all the surface operations, this one appears to have the most skilled labour force, although not necessarily in the sense of formal training. Nevertheless, work experience and skill levels of many workers in this plant are so high that some operations have remained unmechanized because the workers produce a higher quality product than mechanized equipment could.

* * * * *

A CCCR manager:

The old refinery is made up of the anode department, which takes the crude copper from the smelter and casts it into anodes; the tank house department, which refines the copper; and the fine shapes department, which casts the copper into whatever shapes the customer wants. We have our maintenance department and our own shops. Superimposed on that is the electrowinning plant. The anode department today is essentially the same as in 1930. The only difference is that instead of receiving cold metal pigs it receives molten metal. But to all intents and purposes, it is the same as it was in 1930, though we have tried to reduce its labour intensiveness with some specific projects. The fine shapes department has been almost totally modernized in the past ten years. The electrowinning department is highly automated. It was supposed to be very non-labour intensive, but it's more labour intensive than intended.

* * * * *

The numbers of workers in the various categories clearly indicate the labour intensiveness of much of the refinery's operation: 610 operators, 239 maintenance and 17 utilities hourly rated employees, and 170 staff. Over the past ten years the proportion of the hourly labour force in maintenance has increased slightly from one-fifth to one-quarter. In the new electrowinning division, 38 per cent of the 112 employees are in maintenance. Neither ratio (each about 2.61:1, operators to maintenance) suggests much automation.

When the CCCR changed its furnaces from coal to gas, a clean-up crew of five or six people was eliminated and the amount of operation and clean-up work was reduced over all by about fifty people. One CCCR worker says: "They didn't actually lay off people because they could have had tremendous battles in the next contract about automation and technology. If they were actually laying people off as a result of bringing in gas, it would have meant sabotage or something or tough bargaining or strikes over those issues, so they didn't actually lay off guys, and that is why they got it so easy." Inco appears to have a practice of using only massive layoffs rather than smaller ones, usually allowing attrition to pick up the slack. This strategy defuses the introduction of technology as a major point of conflict.

The electro-refining operation accounts for 95 per cent of the plant's copper output and uses a process very similar to that at PCNR and the Thompson Refinery. Copper concentrate is received and enters an oxygen flash smelter, which produces a copper matte in liquid form. The matte is further refined in a converter producing blister copper. The molten blister copper enters a large reverberating furnace where greenwood trees are placed to remove oxygen and help in slag formation. A manual skimming technique is used in the anode furnace. The molten copper is skimmed by a man "raking" the copper into a ladle; he is protected by a shield. Other men attach hooks from the crane to the ladle and operate the crane to move the material to the anode casting area.

Anodes are cast by a highly mechanized method, although a man must judge the proper thickness of the mould on each cast. The metal is poured into a rotating horizontal casting wheel, and the anodes weigh about 580 pounds. The method used here is more efficient than PCNR's casting wheel. After being rotated and removed to cooling tanks, the anodes are lifted by a crane to waiting push carts to be transported to the tank room. There they are suspended in an electrolytic solution and interwoven with cathodes made of thin starter sheets of pure copper. As in the electrolytic nickel refinery, a current passing through the solution causes the copper to move from the anodes to the cathodes. These become the highly refined copper to be melted and cast into products. Achieving this result has been described as almost witchcraft. The quality of the copper is determined by detailed manual work. It is described thus by a plant official:

> *The anodes are loaded into the cells by cranes, but before they can be they have to be dressed – fins have to be chipped off and bent lugs fixed. That's manual chipping. They are put into the cells and every single lug is pounded with a hammer to be sure the anodes hang truly in the cell. When you recognize that there's only 3¾ inches between the anodes and that the starting sheet is hung between them, the starting sheets must be flat and straight and they must be put in by hand. After they're all in, people go around and align them, and the power is switched on; and they go around by hand to check the current – all manual and on top of the cells at 115°F. They will be checked over a two-week period for shorts. When the copper is pulled from the cells it should be well washed, again manually. Any bent corners or nodules with slime should be individually chipped off, manually.*

The rectangular tanks that receive the anodes and cathodes are made of concrete and lined with lead. The 580-pound anodes remain in them for twenty-eight days, but cathodes are produced after fourteen days, each weighing about 250 pounds. The spent anodes, averaging about seventy pounds, are returned for remelting in the anode furnace.

Crucial to the entire process is the production of good-quality starting strips. These are produced at a rate of 2,900 sheets a day by a stripper gang of nine men who earn a bonus of 30 to 35 per cent and can go home when they complete the required number of sheets (they usually arrive at 6 A.M. and leave at noon). They work exceptionally hard and as a team, moving continually, removing strips of copper from stainless steel plates and piling them up for the looping machine. The men on the crew tend to be in their late forties or fifties and work with great precision. According to an Inco official,

> *our sheets-per-hour production is the highest in the world. There is equipment on the market to do this, and eventually, when this particular group of people has passed through, we will have to move to an automated method of stripping. While the equipment is more sensitive, the productivity is no better than we currently have, but we recognize that when this kind of person is no longer available we will have no choice but to automate.*
>
> *I suppose that I don't think I would want to physically work that hard, and I think it only right and proper that it become mechanized, except that I know that is a fantastically motivated group and they are pulling in a hefty buck. Those guys earn their money.*

The strips move to the looping machine; running it is also a bonus job, averaging 28 to 36 per cent. One man feeds the sheets, which are flattened; another feeds the cross rods that support the top of the sheets into the machine; another makes sure nothing becomes jammed; another adds the loops that are used to suspend the sheets; and another takes the sheets off at the end. This once manual job has now become quite mechanized, even though the men still control the machine's output. Tank charging is another incentive job in the plant. It involves removing and replacing anodes once a month and cathodes twice a month progressively from tank to tank. The bonus for this job averages 30 per cent.

After the copper has been recovered, the cathodes are melted and cast into a variety of shapes. Although men still control the amount of copper poured, casting is now a mechanized continuous process. The metal flows into moulds to produce wire bars, billets (rods of copper), or copper cakes (slabs). The billets can be cut to customer specification in lengths up to twenty-four feet. Most of the people in this "fine shapes" section are inspectors for quality control; others transport the product on fork-lift trucks.

Also part of CCCR is the silver refinery, which processes some of the precious metal by-products such as gold and silver. It also was constructed in 1930. About sixty men are employed here. Security is tight; each man is strip-searched when leaving. All refuse, including garbage from the lunch room, is routinely burned to retrieve metal. Much of the work is done by "bucket chemistry," that is, small-scale batches. Once a work place where men were proud to work, the silver refinery has suffered a decline recently. This is partially because of a problem known as "garlic breath," an odor the men acquire because some of the chemicals used enter their bodies. Another problem, even more serious, is a high rate of lung cancer and respiratory disease. According to a union official, of eight employees who worked in the gold room of the silver building for over twenty years (only two work there at any time), four have died of lung cancer and the other four are on disability pensions because of breathing problems.

In 1973 an electrowinning process was added to the traditional electro-refining operations, but it has increased capacity by only 25 million pounds a year. It is essentially a separate method and does not yet produce material of the quality of electro-refining, with copper content varying from 89 to 97 per cent compared to 99 per cent. Electrowinning relies on carbonyl residues from CCNR, which are transported by pipeline from that plant. It processes this material into copper cathodes and precious metal concentrate. Seven men work in its tank house (it was originally designed for two), and in the leaching operation there is a central control-room operator in addition to two supervisors and eight men a shift to, as a manager says, "unplug lines, check to see pumps are running, sit in the lunch room. The same as the CCNR, a policing situation – check for leaks, cleaning up spills." Because the materials processed are entirely in a slurry solution, and a number of tasks necessary in electro-refining such as loading anodes and mudding cells (a cleaning of tanks needed only once a year in electrowinning instead of once a month) are eliminated or reduced, there is not such intensive labour. But the electrowinning process does create technical problems and requires about twice the electrical charge used in electro-refining, with resulting higher costs. Moreover, the titanium cathodes used to form growth deposits of copper about doubled in cost in five years. All this has caused the process to be nicknamed "electro-losing."

In the electrowinning tank house two innovations minimize the acid fumes given off by the electrolyte: two huge fans move the air, and foam chips spread on the surface of the liquid reduce evaporation. Still, the atmosphere contains a good deal of gas, and the tank room is a hot place to work. An overhead crane is used to lift cathodes, and as this method leaves little residue in the tanks, the amount of cleaning is reduced. A stripping machine to remove the copper starter sheets has been used, but men are now doing it again by hand because a higher quality can be maintained.

The support side of the electrowinning method is a highly mechanized battery of pumps and equipment involving fourteen separate processes on a single line. Each of these is divided into operations carried out by an operator and his helper. These men are highly skilled, an operator requiring two years of training and an assistant operator six weeks. They are rotated between the processes so that they can learn all parts of the method. They receive direction from a supervisor in a central control room who oversees the process.

For the first six months after the electrowinning process was installed there was a labour honeymoon with virtually no grievances and low absenteeism, but the next three years were troubled. There was an average of sixty grievances for each hundred men during the problem years. Human relations techniques similar to those at CCNR were initially used (see page 205), but unlike the CCNR workers who were drawn from many locales, most of the men came from the old CCCR tank house; they were more cohesive than the nickel refinery men and retained many of the traditional understandings from the tank house. In the early years the process did not run smoothly, and technical problems resulted in hard work and spills, causing the men to do many "shit jobs." These have now been reduced.

Copper Cliff Nickel Refinery

An atmospheric-pressure carbonyl process of refining has been used since 1902 in Clydach, Wales. Inco began to make experiments with high-pressure carbonyl extraction in 1957 to obtain more effective and rapid extraction using cruder raw materials. Construction of CCNR, the world's first high-pressure carbonyl nickel refinery, began in 1969 and was completed in 1973 at a cost of $140 million.

CCNR has two operating divisions, the Nickel Refinery Converter Department (NRC), which grinds the matte into granulated metallics, and Inco Pressure Carbonyl (IPC), which extracts the nickel into pellets and powder. The plant operates at a thousand pounds pressure per square inch and has sixty-six miles of piping, ranging in diameter from one to thirty inches, and 15,000 valves. Carbonyl gas is very toxic. Because of the danger of leaks occasioned by the high pressure and the number of valves, the company rules that women may not work in the IPC plant (women of child-bearing age are apparently unable to accept the antidote for exposure to carbonyl gas). In the first three years of operation some 130 workers were reported to have been treated for exposure to nickel carbonyl gas. The company maintains a four-bed miniature hospital on the grounds and performs monthly urinanalysis for the workers. The nature of the highly contained process means there is a decline in day-to-day accidents, but the presence of highly toxic gas under high pressure increases dramatically the possibility of massive disaster, including leaks into the atmosphere.

The plant is very much automated. For every operator there are 1.2 maintenance people and 0.9 staff. First of all, everyone had to be trained, and CCNR was one of the first Inco plants to use the modular training system. An employee needs eight weeks of initial training and eight weeks of specific training to be classed as an assistant operator. After a year's experience he is eligible to bid on an operator's job, should one become available. The plant operates on rotating shifts seven days a week except for maintenance, which has a regular day shift. At the beginning recruitment for the plant was very selective. All workers were transferred from other operations and subjected to emotional, psychological, and aptitude tests. It was Inco's first large-scale application of "people technology." Through a variety of group sessions management tried to instil the impression of "one big happy family." Workers were told that the organization was to be "flattened" so that they could have more say. As it turned out, this idea remained in the heads rather than the practices of management. What was delegated was responsibility; control stayed centralized at the top.

* * * * *

One of the original CCNR employees:
Our personnel man would interview us and send us for an interview over here [CCNR], *and they would pick us if we passed the interview; then we had a test that we had to write which was not an I.Q. test but along those lines, and we had to get a certain mark in order to pass. When we first came over, there were about twenty operators, and we were put through a week of indoctrination in the plant, just getting familiar with our surroundings and with the people in the nickel refinery. There were only staff people here before us. After spending a few months on day shift they put us on shift, and we continued to train on this MTS system which gave us book-work training and out into the plant for some on-hand equipment training. We were testing lines and getting the plant ready.*

* * * * *

Right after start-up employees filed over a thousand grievances, mainly because operators were being asked to do maintenance work. Management paid little heed to the existing contractual agreements with the unions.

* * * * *

An Inco official:
We wanted the people in there to be able to use tools to go about their operating functions in the plant, clearing blockages, putting in purges which are used to clean out process equipment, so that they can be inspected and worked on. To do this, if we had to rely on maintenance we would be calling them all the time to do these jobs. There is more continuity if the people can go ahead and follow through the job to a certain extent. It's gone a lot farther than in a lot of the other Inco plants.

The resistance was mainly around the fact that maintenance people are more highly paid than operators, is that correct?

Yes, that is correct. There was a real split in the operating ranks; some of the fellows said they wouldn't mind doing the work so long as they were paid the money, and there were others who said that they didn't want to do the work whether they got paid the money or not. They were going on the principle of the thing, saying, "That is not our work."

How was it resolved?

There was such a mass of grievances that they weren't all handled individually in the end. What happened was that there was a "direct difference" between the company and the union, and that sets up an arbitrator and the decision is made by someone outside. A solution was reached whereby the operating people would clear blockages in their equipment – they would use their tools to do this – but there was a distinction made, and it was a stand-off more than anything else. Operators are allowed to do a certain amount of mechanical work but not other things. Maintenance can do all the mechanical work including the things that the operators are allowed to do, which we needed because we don't always have operators at hand to do the work that they are allowed to do, and if we have maintenance people available we get them to do it.

Do operators now get a little higher rate than they would have if they weren't doing some maintenance work?

No, they don't.

* * * * *

A union official:

When they opened the carbonyl plant they did a con job. They told everybody that this was the most modern place in the world, that you really had to be somebody to work there. And they were going to select the best people for the job. They were going to train them, and they had a fantastic training program laid out to give them. They wanted applications, they didn't want job bids. They would select the applicants, the most suitable people, for this very important job – aptitude tests and everything. And lo and behold, a hell of a lot of stewards got selected. You'd think they would want to keep stewards away, but no. They got the grand treatment. All this isn't old Inco anymore. We're here to work together. Everything is going to run smoothly. You tell us your problems, and we will sort them out. At first we had no grievances from that place, but gradually they began to come in; and we started hitting them with health and safety, and that place went crazy. The big thing was their piling more and more work on the operators. They wanted them to be mechanics as well.

* * * * *

The immediate impression one receives on entering the plant is how few people there are. Although there are 346 employees spread out over the four shifts, they seem almost invisible. There is a fairly large clerical staff by Inco's standards, and some employees work in the hospital and supply rooms; but in the plant itself there are few people. The following is a description by a staff supervisor of the various areas of the plant and the division of labour:

Each area is split up into two sections, and there is an operator in each section. He controls all the operating equipment in his area. On the day shift he spends a lot of time signing in maintenance people, including mechanical, electrical, instrumentation, and checks equipment out for them because carbonyl is a toxic substance and lets them know when an area is safe to work in. He also has with him two assistant operators. They all work pretty much together, but he has control over these assistant operators.

What activities do you normally perform?

Let's talk about day shift. Day shift is the busiest time in the plant because all the other staff functions are going on as well. I would come in at 7:30 in the morning, and the graveyard-shift foreman would tell me what basically had taken place on his shift and what I was going to have to follow up on to keep the plant going. After he is all through with his hand-over and I understand to my satisfaction, he goes home and then the shift is mine. That normally takes fifteen minutes. The mechanical foreman will call me or I will call him at about a quarter to eight every morning and make sure that the jobs we set up the day before for the following day can be followed through and whether there are any changes. My men go out and sign maintenance people into the field for the different jobs, routine chores, leak checking.

What do the hourly rated people normally do?

Hourly rated workers arrive near the same time as I do to relieve their people in the field. There is no punch clock for them; there is just a set of time cards which we sign, and the system we have set up is that when a man is relieved, he goes home, and the next man takes over. Rather than working a strict eight to four, four to twelve, twelve to eight, it is more a half an hour before sort of thing. So he has a pretty good idea even before I go and talk to him of what's happened on the shift before, and he has a pretty good idea of where we are going for the shift. He has probably already made a pass around the area to make sure that everything is okay at the start. So then we sit down and I will give him a rundown on what maintenance, mechanical, instrumentation, electrical work is to be done in that particular area that day and any problems we might expect. I'll do that with both operators, and what I generally like to do, unless I have a specific job for the assistant operators, is let the operators use themselves and the two assistant operators associated with them in whatever way they see fit.

How often in a shift would you see an operator?

We are in constant contact with them because the operators wear one of these radios, as the foremen do, and we are able to reach them at any particular time. We can contact them through radio or a plant-wide intercom system. But I would say, except for two hours of the day, you are in constant contact with them.

The heart of the plant is the main control room, with a bank of panels that keeps tabs on the entire operation. It is highly automated, and from it the control-room operator has radio contact with people in the plant. Although he is at the centre of the operation, he is basically responding to instruments and has little direct human contact. The job is, as can be imagined, very tedious. One supervisor said:

It's a very responsible position, but I don't think that any one person should be on that job for any longer than a specific amount of time. . . . There are over 6,500 different pieces of equipment he would be controlling, and there has got to be a specific amount of time that a guy should spend in there, because with a room such as this, the only time he can leave during the day really is to go to the washroom. He is there the rest of the time. He even eats lunch there.

The only non-automated activity in the plant is the packaging of powder. It is the one operation over which the worker has some control. He manually operates a lever putting nickel powder into three-hundred-pound drums, puts on the caps, and writes the batch numbers on the drums. This job will be eliminated in two or three years when the plant switches completely to making pellets. The pellet packaging area is already automated, and the work is primarily machine watching. This is the plant's primary packaging operation with the loading of five-hundred-pound drums and some bulk handling of nickel pellets into rail cars for other plants.

* * * * *

A CCNR supervisor, asked about controls against sabotage:
There have been times when we thought we had sabotage, but there has never been anything provable. It is very, very difficult to control against sabotage. If there is any, if there ever would be any, it is a very difficult thing to

control. As you noticed when you were in the plant, you can go through quite an area sometimes without seeing a person. That can happen quite regularly, especially on shift. If there was someone out there that wanted to do some damage, by knowing the systems and turning the right valves perhaps he could do a substantial amount of damage. But he would probably be risking himself if he was ever to try and do anything dangerous. We have had contamination in our product; a customer has sent us back powder with cigarette butts and Dustbane, pop cans, and this sort of thing. It shouldn't be like that.

* * * * *

Automation has reduced flexibility at CCNR. In order to cut back the labour force it is necessary to close down the entire plant. Even when it is operating at only 65 per cent of capacity, as it was in the summer of 1977, it still requires the full complement of workers. And whether the plant operates or not, the capital – $140 million – remains tied up in the equipment. As a means to reduce labour costs generally the plant has been very effective. It has allowed Inco management to mothball PCNR's electrolytic operation. Yet as an experiment in industrial relations the plant has not been totally successful; there is a good deal of resentment between workers and foremen and top management over divisions of tasks and lines of authority. As far as the company is concerned, the experiment in training technology has proven successful enough to expand it to the entire Ontario division. It has given the company its first major breakthrough in the operator-maintenance distinction, a breakthrough that is likely to be exploited in other operations (see page 288).

The most obvious danger in the plant is the high-pressure carbonyl gas. Set up as the plant is, leaks of gas under a thousand pounds of pressure per square inch are likely. Even more serious would be a major escape of gas that could affect the entire community. The basic work for operators is patrol, keeping watch over the very large equipment areas and taking instruction by radio from the control room. The maintenance workers, like those in most automated plants, are essentially trouble-shooters. They make repairs to the equipment, but because of the particular hazard of carbonyl gas they must be constantly on guard.

Even though a great deal of automation has been introduced into refining – all of CCNR, parts of both CCCR and Thompson Refinery – and Inco's most labour-intensive operation has been moth-balled, many aspects of the process still demand skilled craftsmen-operators. Science has not completely replaced art (with the notable exception of CCNR), but the direction of Inco's strategy is clear: use capital to replace labour wherever possible. This means the amount of bull-work is reduced, but it also means, as it did underground and in the mills, a decline in the number of skilled operators. The essential complement to this strategy is a change in the way skills are transmitted. Rather than rely on apprenticeships, formal or informal, where workers train each other, the company has adopted "scientific" training in which the equipment rather than the person is the focal point for organizing production. Management appropriates, codifies, and reformulates these skills into standardized procedures that it can rapidly transmit to new workers.

Only a few key surface jobs are on bonus, such as the copper stripping at CCCR; otherwise, work is straight wage labour, although PCNR has unofficial production quotas. Patrol work with automated equipment or machine-tending where the rate is set for the worker requires little monetary incentive to maintain productivity because workers cannot control the pace of work. The bonus jobs that remain on surface are anomalies, and time will see them eliminated.

Not all capitalization has its intended effects. The electrowinning department of CCCR is a good example. It is highly automated and was designed to produce high-quality copper with little labour. Neither goal has been achieved. Its quality is low; in several cases it has had to revert to traditional labour-intensive methods, and closely knit tank-house crews were resistant to the change. CCNR, on the other hand, has performed as expected, although its workers have offered some resistance to attempts to destroy the traditional distinction between operators and maintenance men. At CCNR management was eventually able to overcome some of this resistance because it was able to force new agreements with the workers.

The Technical Operations

Sampling is a technical process that takes place in all surface operations and is central to the successful production of uniform quality nickel and copper. It is not organized the same way in all Inco plants, however. In the Ontario operations each plant has its own group of samplers; in the Thompson complex the samplers are organized as a separate section. Another aspect of the technical operations is research and development. Most of this work is done at Sheridan Park and is oriented to improving methods of milling, smelting, and refining the ore. One type of research and development known as process technology takes place within each of the plants and is directed toward trouble-shooting and making relatively minor modifications to the plant operations.

In this field, the central research laboratory tends to act as the source of highly trained professionals for the dispersed process technology operations rather than the other way around. This tendency is becoming more marked as Sheridan Park matures (it has now been operating for over ten years). At the laboratory there are very clearly defined divisions between the forty-eight professionals, the sixty-six technicians, and the twenty support workers. Nearly all the professionals have doctoral degrees as chemical engineers, metallurgical engineers, or chemists. Of all Inco employees, aside from management, they have the best working conditions and greatest innovative potential, although even their work is controlled.

They are given very demanding work-loads. One Sheridan Park professional reported: "The first day I was doing four or five projects, some of which were long term, some short term; some were just along the laboratory scale and some were working with either our pilot plants or our production facilities. It was a relatively broad spectrum." The professionals report to their section heads once a month but normally see them daily. The section head establishes priorities and screens projects, but the professional has a good deal of autonomy: "If I have six projects, I am the only one who will say, 'Well, I'll get three people on this and two on this.' " While the official hours are from 9 A.M. to 5 P.M., most professionals arrive around 8 A.M. and do not leave until after 6 P.M. According to one professional, "They love their science and they really enjoy what they are doing; the stimulation comes from the work they're doing and also from management." Additional motivation comes from the monthly review sessions, where each professional presents his projects and findings to management for scrutiny.

Working under the professionals are technicians. Although clearly directed by the professionals, they are also given a fair amount of autonomy.

* * * * *

An interview with a Sheridan Park professional:
What is your relationship with the technicians? Do you give them specific orders or direction or just say, "Do the following experiment," and wait for the results?

I find it better to explain to them as much as possible why we are doing this and then I indicate what type of data we are looking for. . . . They perform the experiment. I indicate, usually, the conditions. The technicians finish the experiment and submit the samples and wait for the data to come back. The better ones, of course, actually partially analyse the data and the calculations. The ones who are not as competent would just present the basic raw data.

* * * * *

This low level of supervision is confirmed by a research technician: "I get to make a lot of, well, not decisions – they're not really earth-shaking decisions – but I get to follow through on everything that we are doing. I have my own say-so. If he says this and I think that, then we will do both of them, or he will take my idea into consideration." This person says her work has great variety and could not be called boring. Her training is a diploma course at a community college which provided her with the basic skills to begin work immediately in the laboratory. Because of this background she is limited in her career mobility. If she wanted to make a major advancement she would have to go back to school and get a university degree; otherwise she will remain a technician. Unlike the professionals, she comes to work at 9 A.M. and leaves at 5 P.M., although she says it is good for appearances to arrive a little early and occasionally stay longer.

The basic task of the technician is to do experiments as directed and keep laboratory books about all the particulars. As she describes it: "Technicians are supposed to and usually do all the practical lab work and all the analysis. The Ph.D.s get their results and write up the report and submit it to their section heads, and the section heads say, 'Well, that's a useless project – throw it out the window,' or 'That's a good one; we'll work on that for another couple of months.' " Generally the pace of work is reasonable, although there is a good deal of competition, and pressure mounts before the monthly report meetings: "At the top, where the guys are the Ph.D.s, it's very competitive: 'My project's more important,' 'Mine's more important,' and 'Hurry up and get this done, I've got to beat him out for this.' We feel the pressure a lot. Once every month they have a planning meeting where everybody has to go, and all the bosses give a presentation about what they have been doing for the past month and what they propose to do in the next month. That's when the pressure is on."

The hierarchy within the laboratory makes a clear descent from the managers to the section heads to the professionals to the technicians. To some extent the professionals resemble the supervisors in the other operations except that they in fact work; like supervisors, they are subject to pressures from above, yet they are not under much constraint from workers because giving direction to others is only a minor part of their role. They often work along with the technicians rather than simply direct them. The workers have the products of their labour appropriated by management; the knowledge of the professionals is appropriated in the same way. Credit for discoveries and ideas is given to the section heads rather than to the professionals: "You sign [an agreement] that anything you discover belongs to Inco." They have limited autonomy but not control over what they produce; to some extent they give direction, but they also receive it. Although clearly in a privileged position with the company and highly rewarded, the professionals need to become section heads before they are part of management.

The work of process technology employees in many of the surface operations has already been discussed. It is easiest to study their actual organization in Thompson, where they form a distinct division. There are 150 people assigned to this division, including 40 samplers and 35 engineers. The laboratory operates on a three-shift, seven-day basis, although most of the work is done on day shift. Thirty-eight people work directly in the laboratories, including chemical analysts, section leaders, and supervisors and their assistants. As at Sheridan Park, many of the process technology workers are technicians. They are recruited from community colleges and their hierarchical position is similar to those already discussed. At the senior levels the professionals resemble those at Sheridan Park, although "in the field" they are closer to the actual implementation of new designs or testing of various processes.

The samplers are in a different category, as they are hourly rated rather than salaried like the others. They are divided into junior and senior classifications; somewhat like stope leaders, the senior samplers have a little more authority to ensure that work is done properly, but they still do a good deal of the work themselves. Samplers have no formal training. They learn the routine by observing samplers at work for a few weeks. The job involves taking samples at various product stages and work locations – dust, filter cake, slurry, or crushed ore – at one of twenty-five sample points, placing them in containers, and returning them to the sample house. Samplers do little analysis themselves, only preparing the samples for the laboratories where the analysis is actually done. Some items are sampled each hour, some each shift, others daily, and some monthly. The supervision on the job is not very close, but one sampler said, "I couldn't really cheat on the job because I have to get the samples in on time and get them done."

Virtually every operation in the plant is sampled. There is a check-point for everything going out – finished nickel, copper, waste, even the garbage – and periodically the yard is sampled. Over the past seven years the numbers of samplers has declined from sixteen on three shifts to nine on four shifts in Thompson. This was accomplished by doubling up on jobs and having the operators do some of the sampling so that the samplers merely pick up the samples at the end of the shift.

As their job keeps the samplers moving throughout the plants, they encounter heat, dust, or gas in a variety of environments. One sampler said of her job:

> *As far as the work being physically heavy, there is some, but the majority is not; but most of it has to do with your lungs. The other day someone was saying we are overpaid, and I said, "Really! You can't put a monetary value on my lungs, thank you very much." And they looked at me, and I said, "We are not slugging the muck like the miners, but how many times do you wear a dust mask? Just check out what we are doing." The big thing is dust. It's silica or asbestos constantly. We are so mobile, we are exposed to everything that any worker on surface is.*

All surface jobs are exposed to a variety of unhealthy elements. The severity varies with the location, but few locations are not subject to some kind of unnatural conditions. The technical operations employees, but particularly samplers, experience most of the conditions other surface workers face. They are directed by shift bosses and have to perform certain tasks on time. They are also subject to the consequences of mechanization and automation; many of their tasks are being absorbed by machines, as at Copper Cliff Mill.

Technical workers at Inco occupy an ambivalent position between management and labour. Except for the samplers, all are salaried workers and not part of a union (although technical workers at Falconbridge are members of United Steel Workers of America Local 6185, and the USWA has made two unsuccessful attempts to organize technical workers at Inco, the most recent in the early 1970s). On the other hand, all professional and technical workers below the level of section head are subject to the direction of section heads – the level at which management begins. The nature of their work demands that they be given a degree of autonomy in the performance of their work, but limits are set and the products of their work are appropriated by management. Technicians are directed by professionals much as many hourly workers underground and on surface direct other hourly workers. Among technical workers, however, there is a clear distinction between the career paths open to technicians and to professionals, whereas it is possible for most of the hourly rated workers either to bid for training or to wait for seniority and get into a classification that directs other workers. Technicians are, in fact, trapped in their classification, regardless of their experience and skill levels, unless they move outside the organization and acquire other certification.

Underground, the tendency is toward more detail labour as a result of mechanization, so that miners are becoming machine tenders or machine operators. On surface, the tendency is away from machine tending and toward patrol work as a result of automation. Underground, management is shifting from a "responsible autonomy" strategy, under which workers organize production themselves and are induced to maximize production by the bonus system, to a "direct control" strategy under which workers are concentrated into fewer work places, subject to a detailed division of labour, and have their work supervised more closely. Direct control strategies have always been in force on surface. With relatively few exceptions, the work process has been closely watched by supervisors, and workers have performed a limited range of detailed tasks.

Management has now chosen, to the extent possible, to minimize the number of workers. It has instituted experiments in human relations techniques of social control for the remaining workers, but these have been limited in their impact. In an industry where worker-management conflict cuts deep it is not an easy matter to develop human relations techniques. Traditionally the company has used a stiff penalty system to regulate the work place and layoffs to regulate the work force. Faced with the cyclical nature of labour shortages, a highly cohesive union, demands for safer work conditions, and high wage demands for skilled workers, Inco management has introduced a training scheme that will give it more flexibility in meeting its labour demands and reduce its reliance on skilled craftsmen. This is the most recent strategy of containment: create conditions in which unskilled labour can be brought into the organization and quickly trained through standardized methods to operate or maintain a number of specified pieces of equipment, thus reducing the need for experienced or skilled workers.

It has become evident from changes occurring in both underground and surface work that management has introduced technology to increase its control over production. Technology has been used in such a way that it destroys the traditional ways of organizing production with which miners and many surface workers have controlled the direction and pace of their work.

CHAPTER SEVEN

Health and Safety in Mining

Even if a guy does everything right, he can still get killed. That's one of the hazards of working underground; you can get killed even if you're innocent.

A Thompson miner

No one can enter a mine without a sense of danger. It is an unnatural experience and the environment is unfriendly. Every miner knows stories of fatal accidents or has had friends killed.

* * * * *

A Thompson miner:
We think as miners that it is a job not like other jobs. I've worked in different jobs, and there's no way it's comparable to any job I ever worked at before. I mean, you go down there and you may not come back out. Every day you say to your wife, "I'll see you tonight," but you may not see her tonight because that's mining. It's dirty; it's wet. Maybe it takes certain guys to go into mines; I don't know. Are they stupid or what? Where is there any job that is so restricted in how you do your work? You've got the Mines Act, and you are responsible for all the guys who are in your stope. . . .

I will tell you about the accident. It was a Friday night right at quitting time. The guy that died, he was working at the other end of the stope, timbering. We were working in the muck pile, me and my partner, my driller. The guy came up to check and see if we needed any caps for the blast, which is the normal thing to do. The guy had a couple of years' experience in the mine; he was no fool. He had been around. He was twenty-one. I worked with him in a couple of stopes. He crawled under the hanging breast to see, and we were knocking bits off our drills. And my driller went out and spoke to him and asked him what he wanted, and he said, "Do you need a guard?" and the driller said, "No, but maybe you can go and cut that chunk that's hanging on the bolt down on the hanging breast." So the guy goes over and he is cutting a chunk of rock off with a hacksaw. There was another chunk further under the breast, so the guy went to get that one, and the bolt where he'd cut and the ground above it was loose. He was reaching under to cut the other one and this piece fell. It slipped down off the bolt and fell on top of his back. He was not a hefty guy and one leg was out here and the other one was back this way. It never dropped from much height. It maybe dropped a couple of inches onto his back, but still it was solid rock. The thing must have weighed tons. It took nine guys to get it off him. We had no big timbers to pry it off him, and then it was too late. There was not a bone broken or anything. He was all squished.

The Mounties were down, investigating, and everyone else. We were on afternoon shift then, and it was a Friday night. We were there until about four in the morning, waiting around and answering questions. Then a month later an inquest came up and we all just more or less said what we were doing at the time, and there were five or six guys involved in it. The stope crew plus the guys that came in and helped plus the shift boss. Well, the thing is the way the judge said it: "Under the circumstances of the death of so and so, I find the following persons guilty of negligence. No person or corporation is guilty of negligence." But I just knew that everybody, even myself, their heart was jumping out of their chest. It was just a freak accident.

It must create a tense atmosphere in the mine for the next while after a major injury or death.

Everybody's quiet. The supervisor starts running around looking for minor things. They don't pick up the major thing that caused the accident; they pick up stupid things.

So they are afraid as well because their neck is on the line?

The Mines Act is law, and they are afraid that a mine inspector may come down and get them for a little thing.

Can you, individually, be held responsible?

Just say I'm guarding [standing watch during a blast] *and somehow someone gets by me and they get killed. I'm up for manslaughter. Inco's not up for manslaughter; I'm up for manslaughter, personally. The mine captain pointed that out to me; he said, "You have got to do this and you have got to do that and you have got to watch the operator and check the walls; and if something happens in the stope, you are responsible." They pretty well put the blame on your shoulders. . . . Mining is a dangerous job. I don't care what anybody says: once you get on that cage and go underground, you are a potential casualty.*

* * * * *

Danger in Mining

The mining industry is the most dangerous in the country, rivalled only by forestry in this unenviable position. Table 20 shows the distribution by industry of occupational injuries and illnesses that contributed to work-related deaths in 1975. Although it employed only 1.4 per cent of the labour force, mining accounted for 6.9 per cent of the deaths due to injuries and 57.5 per cent of those related to occupational illnesses. It is evident from Appendix X that the incidence of death arising from mining between 1967 and 1976 was more than ten times higher than the rate in manufacturing. Over that decade some 1,670 people were killed in mining. Moreover, the rate has not improved since the Second World War; the fatality rate for

Table 20

Fatalities in Canadian Industry Caused by Occupational Injuries and Illnesses, 1975

Industry	Occupational injuries		Occupational illnesses		Total fatalities		Per cent of workforce
	No.	%	No.	%	No.	%	
Mining	70	6.9	88	57.5	158	13.5	1.4
Manufacturing	176	17.3	46	30.1	222	19.0	21.0
Construction	209	20.6	8	5.2	217	18.6	6.5
Transportation	212	20.9	4	2.6	216	18.5	8.7
Trade	71	7.0	3	2.0	74	6.3	17.5
Service	82	8.1	1	0.6	83	7.1	26.9
Public Administration	81	8.0	3	2.0	84	7.2	6.9
Agriculture	13	1.3	...	...	13	1.1	5.2
Forestry	71	7.0	...	...	71	6.1	0.8
Fishing	27	2.7	...	...	27	2.3	0.2
Finance	3	0.3	...	...	3	0.3	4.9
Total	1,015	100	153	100	1,168	100	100

SOURCE: Calculated from *Labour Gazette* 77, no. 12 (1977): 557-58, Tables 2 and 3.

mines in Ontario in 1976 was identical to that in 1951, 0.39 per million hours worked.[1]

Appendix XI shows the fatal injuries and compensable injuries in Ontario mines between 1970 and 1975. A marked trend in which the fatality rate tends to decrease while the injury rate increases is clear, and so far there is no satisfactory answer to the appearance of this pattern. This is, however, the period during which a good deal of mechanization was introduced, and it may be that mechanization will produce more injuries as a result of the greater amount of machinery underground while at the same time reducing fatalities because the miners work less under loose rock. The Ontario Royal Commission on Health and Safety of Workers in Mines (the Ham Commission) notes that Inco follows the pattern of the industry in this regard: "Unexpectedly, there is little correlation between fatality experience and non-fatal injury experience. For example, in Inco Ltd. whose employment is about half that of the industry, the non-fatal injury frequency has been rising significantly while fatalities are below average for the industry."[2]

Table 21

Frequency of Fatalities and Injuries in Inco, the Nickel Industry, and All Mining, per Million Man-hours, 1970 to 1974

Fatalities	Inco	Nickel	Mining
Reduction	.013	.035	.045
Shops	.183	.094	.066
Open pit	.000	.000	.160
Underground	.301	.308	.446
All	.152	.162	.217
Injuries (compensable)			
Reduction	47.0	44.0	16.5
Shops	17.1	15.1	28.4
Open pit	36.1	36.1	41.2
Underground	90.9	80.0	66.9
All	62.9	54.8	44.9

SOURCE: Calculated from Ham Commission *Report*, pp. 120, 130, 140, Appendices D.5 and D.7.

Table 21 demonstrates some differences in safety levels between Inco, the nickel industry as a whole, and the mining industry as a whole. Nickel in general, and Inco in particular, had lower fatality rates between 1970 and 1974 than the mining industry as a whole, but during the same period Inco's injury rate was higher than the nickel industry rate, which in turn was higher than the rate of mining as a whole. The main differences appear in the reduction plants (mills and smelters) and underground, where Inco's rates are higher for injuries and lower for fatalities. In order to gain a comparative perspective, we should note that the manufacturing sectors during the same period had a fatality rate of 0.033, much lower than any of the three categories examined here. There can be little doubt about the hazardous nature of mining.

Statistics on injuries in mining can be confusing. At one time a lost time accident was recorded only after a worker had missed five days of work, later revised to three days' missed work. Now an injury is reported if anyone misses one day. Injured workers have often been ordered to report for "light duty" work in order to keep the company's compensation payments lower. Ray Moreau, president of Local 6200 USWA in Port Colborne, reported to the Ham Commission in 1975 that there was "still a great deal of hiding injured men in the plant and getting them to perform some simple task." He continued, "The Workmen's Compensation Board, of course, are not informed of these cases and so statistics are never accurate. We only know that a lot more people should be off the job recovering from their injuries than are reported. These walking wounded still receive their wages."[3] The following interview with a Sudbury smelter man and his wife illustrates the reliability of the statistics:

Wife: *He was hit by a thirty-five-pound brick on his foot.*

What job were you on?

Smelter man: *We were tearing some brick out of a converter they put on repairs. We had the big bars pulling down the brick. One of them slid right on my foot.*

Wife: *He had to go to work. He had to walk from the gate to the plant with no help so they wouldn't have to chalk up a lost time accident. They let him sit in the office, but from the gate to the plant he had to make it on his own, even if he crawled.*

Who did that?

Smelter man: *Inco.*

Wife: *It went on for three weeks.*

Smelter man: *Just as long as you showed up and sat there. I was there at 7:30 in the morning and they had a chair and another chair for my foot. They looked after you.*

Wife: *They were trying for an award. If he hadn't stayed, they wouldn't have got it.*

How did you get to work?

Smelter man: *I had a driver. He had a panel truck that would come to the door. I would get in and go to work.*

Since 1972 Inco's official policy has been that workers unable to perform their normal duties are not to report to work (see Appendix XII), although it is evident that this policy is not universally applied.

Injury rates may vary with the type of reporting, but fatalities do not. Death is a fact of life in mining communities, as a Sudbury miner explained: "A dozen [deaths] seems to be the magic figure at Inco. If there were twelve people killed in one year, people kind of look at it as if, well, they were expecting it. If you have fourteen killed in one year, then the people are kind of upset about it." This sacrifice of miners' lives could be avoided. A recent Manitoba government study of fifty-four deaths in that province's mines between 1971 and 1977 found that conditions were "normal" when half the deaths occurred and that workers were following "standard" practices at the time of half the deaths. It concluded that "two-thirds of the deaths could have been prevented with better training. A full 91% of the accidents 'could have been anticipated.' "[4]

A number of factors are associated with fatalities in mines. Fatalities are concentrated among the unskilled, the semi-skilled, and group leaders in mines; 88 per cent of mining fatalities in Ontario between 1965 and 1974 came from these groups. This is well above their proportion of the mining population. First-line supervisors experience a risk rate about equal to their proportion of the population in the mines, but skilled tradesmen have a considerably lower rate. Fatalities are also particularly high among workers with five or fewer years' experience, who account for 62 per cent of fatalities but make up only 26 per cent of the labour force. The shift with the greatest risk is graveyard (11 P.M. to 7 A.M.), especially among those working alone. The principal reasons for fatalities are falling rock (25 per cent), fall of the worker (22 per cent), and accidents with hauling equipment (20 per cent). They account for two-thirds of all mine deaths.

More compensable injuries take place on the afternoon shift (3 P.M. to 11 P.M.) than on the other shifts: 37 per cent of the injuries in only 28 per cent of the time worked. Three-quarters of all mine injuries were caused by slips and falls (27 per cent), collisions and over-exertion (18 per cent each), and being pinned between objects (12 per cent). Injuries tend to occur most often at the beginning of each shift and immediately after lunch break.[5]

The following is a miner's account of four fatalities that occurred in the mid seventies in Inco's Frood Mine in Sudbury:

Within a year's time there were four men killed. Three of those men should never even have been hurt. But the fourth man, there was no telling that the danger was the way it was.

One man was putting up a vent pipe that has a damper inside it to control the air flow. He was working on a scoop bucket, and he slipped and fell and fell to the drift floor about ten feet down; and this damper fell down on top of him and crushed his head. He should never have fallen off there; it was one chance in a thousand that the damper fell on him and killed him.

Another man was in the safest place on the whole level. They had blasted in a place that was being excavated for a new garage on the level, way out in the foot wall where the ground was good and solid, and he climbed out on the muck pile to start scaling. He was a new man; he had only been underground a few months. But instead of following the standard practice of going along the side and scaling the wall and just over his head as he went up, he climbed right up the centre of the muck pile and started to scale standing right out in the middle. And as he was scaling a chunk there, another chunk over his head fell on him. He should never have been hurt. If he had scaled from the side up to the top of the muck pile – scaled from the safe place towards the centre – he would have scaled both [chunks] down and no trouble. He was killed.

Another man, he was standing, watching, on another level. He had a train car full of ore off the track, and one of the ways you have of replacing train cars, especially in the muddy areas (and that was a muddy area), is to take a piece of timber about six or eight feet long, put it against the top of the car and against the frame of the next car or the frame of the engine and run the engine forward, lift the car up, and let it swing over onto the track. We have done that thousands of times. This time, he pushed the car up but it didn't swing over; it just sat there. So this guy grabbed a piece of timber that was close by and went to put it underneath the wheel to hold the wheel up and just as he reached there the car was going alongside of him, and there was a post beside his head. Just as he leaned down between the car and the post, the car swung the other way instead of onto the track and crushed his head. Now that's something that an experienced motor crew always watches for, that they don't get into a narrow spot like that, but he didn't.

And the other case, they knew the ground was bad in that drift and they closed it down the previous fall and left it all winter so nobody would be working in there while there might be water in the cracks loosening it up. Spring came and everything was thawed out, and they were going to scale it and rebolt it. So they sent a scoop driver in there to pile up a bunch of ore, fill the drift half full so the driller could work on top of the muck pile to blast the roof down and rebolt it, rescreen it. He was dumping one bucket, and you always had a habit of shaking the bucket a little bit to shake the last of the ore out. He hit the ore just enough and shook it loose and the darn thing, there were about six or eight bolts holding that big loose. They were just bolted right in the loose, it had broken off so high. Well, he hit that, jarred it and gave it enough strain to start breaking the bolts around it so it broke off about twelve or fifteen bolts, and all this stuff came down. Where he was sitting, he was right close to the edge of it, but the stuff rolled down and crushed him. If he had been two feet further back or two feet to one side, he would have been all right. He might have been hurt, but he wouldn't have been killed. And there was no way of knowing just how bad that drift was. They knew it needed to be fixed up, but that is what he had orders to do; start the work to fix it up.

Why is it so dangerous to work in a mine? A combination of conditions, both natural and man-made are responsible. A statement by the United Steelworkers before the Ham Commission graphically portrays the underground environment:

> Immediately attendant when a miner steps onto the cage and leaves the surface are dangers that include problems related to gravity; immense pressures from underground rock that is frequently described as "solid" but which in

> fact is constantly in varying states of flux and change; total absence of natural light; working with heavy equipment which for the most part is designed for maximum efficiency, and not for the protection of the workers using it; the use of high explosives, high noise levels, air concussions, etc.
>
> Added to this list of hazards are long standing and frequently not understood hygienic, sanitary and health dangers arising from dusts, gases, fog, oils, deep holes, falling rock (loose), runs of muck (broken ores), slippery and unsure footing, and in some instances (as in Elliot Lake) ionizing radiation. On occasion oxygen deficiency and/or fire is another potential fatal hazard.[6]

Underground one's senses are dulled. Two of the most important senses for protection from danger, sight and hearing, are impaired. As Jim Hickey, Compensation and Welfare Officer for Local 6500 USWA, told the Ham Commission:

> Outside of the narrow strip of lighted area, all is dark and unknown and hidden dangers exist. One has only to think of the fears that arise when standing close to the unguarded edge of a tall building – one does not approach to within feet of the edge. In dark, unlighted areas, the senses are not sharpened: but rather, they are dulled by the fact that poor lighting or no lighting lessens the sense of danger and so actions are performed in this environment which would never be attempted in broad daylight or in adequately-lighted areas.[7]

Noise levels are very high near all types of machinery such as drills, scooptrams, slushers. Besides dulling the senses, extreme noise can contribute to permanent loss of hearing. As a result, miners are usually required to wear protective earmuffs. But this protection against deafness further lessens alertness to other immediate dangers, and critical warning time is often lost as a result. Table 22 gives some indication of the noise levels in various locations in mining. Given that any noise level over eighty-five decibels is considered hazardous, all of these areas present risks to the workers, and this is particularly true of underground. In a sample of compensated

Table 22

Noise Levels in Selected Mining Locations

Underground	dB (A) or decibels*
Crusher	110
Drills	115–125
Scraper	115
Open Pit	
Drills	100
Truck (in cab)	95
Tractor (in cab)	108
Surface	
Ball mills	103
Rod mills	96
Copper refinery, vertical furnace	102–110

*"In a noise level of 90 decibels it is difficult to sustain communication with another person by shouting at a distance of about one foot (mouth to ear)" (Ham Commission *Report*, p. 229).

SOURCE: Ham Commission *Report*, p. 227.

hearing cases examined by the Ham Commission, 90 per cent of the individuals had worked principally underground. Ninety per cent had been in the industry fifteen or more years, so that they were between forty and fifty years old when they experienced compensable hearing loss.[8]

* * * * *

A Thompson miner with twelve years in the mines:
The danger is something you're always aware of; even when you go home it still bothers you. It's hard to face in the morning. There's always something in the back of your mind that you might get hurt.

* * * * *

A different kind of factor that may be related to danger underground is the bonus (see page 272). The company denies it; union officials argue that it plays a part (albeit usually in private, because to speak against the bonus means almost certain defeat at the polls for a union official); and the Ham Commission virtually dismissed it with the sole comment: "It has been debated whether or not the bonus system increases the likelihood of accidents."[9]

At a 1976 inquest into the death of a man who had his head crushed during a tramming accident when a car went off the track, one of the miners testified, "He was working on the bonus system at the time of the accident, so he wanted to replace the car as quickly as possible; 'the more muck you pull, the more money you get.'" A second miner speculated that the dead man "decided to help with the car's replacement because it was blocking the track and his own train could not get through."[10] This, of course, does not prove the case but must be considered. Before the Ham Commission, Colin Lambert, on behalf of Local 6500 USWA, had the following to say about the bonus:

> The incentive or bonus system is another area where the company encourages the employee to ignore safety rules in unsafe and unhealthy conditions. This system is designed to give a miner extra money for working at an increased rate. This of course means that he will ignore rules and take shortcuts to earn a few more dollars a day. I personally have approached drillers about unsafe conditions in their work area and have been told by them "there is no money in cleaning up or in scaling loose."[11]

The following day, C. Hews of Inco told the Ham Commission that 42 per cent of fatalities occurred among those on incentive and that 40 per cent of underground workers are on incentive.[12]

Once again a differential between fatalities and injuries appears. In 1976, John Rickaby, Inco's superintendent of administration for safety and plant production, said that 58 per cent of injuries happened to workers on bonus.[13] This figure was considerably higher than Hews's 42 per cent of fatalities among bonus workers. One might conclude that fatalities are not as closely related to bonus as injuries are. There is, however, some additional evidence. In 1976, Tom Gunn, co-chairman of the Steelworkers' Inquest Committee, and Dave Cochran, co-chairman of the Health and Safety Committee, revealed some statistics they had gathered. They maintained that sixty of the eighty-six deaths at Inco since 1960 (70 per cent) were "bonus related" and that "people on bonus" made up "85 per cent of the accidents" at Frood Mine in the first seven months of 1976. In the same article Gunn is reported to have said, "You don't get paid for fixing things. That's called reconditioning and that cuts into bonus time. So they try to get by as much as possible without fixing things and conditions deteriorate. If a bonus miner complains about a safety hazard the mine foreman will shove him the job of repairing it." Cochran is reported to have added: "If a miner followed all the safety rules, even if he knew them all, and the mining regulations, he would no way in hell get bonus. Bonus is the company's way to make miners break the rules."[14] Some underground workers agree from their own experience. A Sudbury miner stated his view this way: "Safety is stressed [by management]. They stress safety a lot. Whether they believe it or not, I really don't know, but they do stress it. But safety's non-existent as far as I can see in the drift. None at all. The fellows there have a job to do, and they do it. If it is an obvious hazard, maybe they would take care of it. But if they could get around it, they'd get around it because they have to make money. There's a special breed of guys in the drift, and whether they're money-hungry or whether they're in debt badly, I really don't know."

Health and safety problems in mining are not confined to underground operations. Industrial accidents tend to be more common underground, but industrial disease, resulting from the work environment, seems to be more prevalent on surface. Fortunately, one type of operation that in the past accounted for the highest incidence of cancer in mining is no longer in existence. This was the sintering process formerly located in Port Colborne and Sudbury. The Copper Cliff Sintering Plant operated from 1948 to 1963. During that time some three hundred men worked there each day, and about one hundred more entered the plant daily for maintenance and service. It is estimated that some fifteen hundred men worked there during the life of the plant. John Gagnon, who worked in the plant from 1951 to 1963, says, "The sintering plant had severe dusty conditions. And because of the fact that nickel dust will not dissolve, but will remain in the lungs, and its abrasiveness will go on irritating the lungs until a tumor is started . . . because nickel is a carcinogenesis, chances are these tumors will become malignant."[15] He has begun an attempt to collect the names of those who worked in the plant, estimating that two hundred have died of cancer at an average age of fifty-two. He has successfully processed thirty-eight claims with the Workmen's Compensation Board.

An Inco employee who worked in the sintering plant in Sudbury for five years recalls that Inco "purposely kept the windows closed and the place closed up because they wanted the nickel dust to settle." Besides the nickel dust there was a great deal of gas in the air from the concentrator. When workers could no longer breathe they would break the windows. A permanent crew of seven carpenters steadily replaced windows as the workers broke them. Every area of the plant was subject to the dust, he recalls, but some areas were worse than others:

> *I remember the baghouse jobs. We had to go into the baghouse, where they had bags like burlap sacks that came from a ring on the ceiling, and the nickel dust went through and was trapped. The mechanical crews would go in every day and take the bags down, put new ones up. You would have to go in with one of these little coal shovels about three inches wide, on your hands and knees with your eyes closed. You couldn't walk because you would fall into the holes in the floor. You felt around to find where the dust was and shovelled it down the hole, held your breath, and ran out on the roof while the next guy went in. Another job was going into the flues that sucked the nickel dust. Dust would build up where the duct work was welded. You had to crawl on your elbows and knees because you couldn't get any higher. You couldn't turn around in the duct, so you crawled up, feeling for build-ups.*

In Port Colborne, former sintering plant workers have filed almost one hundred claims for lung cancer and forty for sinus cancer.[16] By the time this plant closed in 1958, some twenty-four hundred men had worked there. They said that the dust was so heavy the visibility was no more than three feet.

Other surface plants continue to operate under conditions dangerous to the workers' health. The danger is most obvious in the smelters, where workers are frequently burned by molten metal and are subject to high levels of noise and gas. This environment causes tension as well as physical problems. A smelter man told of his own experience: "There's two or three skimmers younger than me that have been laid off. There's one that's been off for three months now. A nervous breakdown. And there's another one there who didn't want to skim. He has an upset stomach. His stomach has been bothering him, nerves. I get that way sometimes. . . . I just make myself sick over it and I just can't go to work that day."

There is evidence that lost time accidents are less likely to occur in surface operations than in underground operations. Appendix XII demonstrates that the three surface operations at Thompson have fewer accidents per employee than any of the mines or even the maintenance department, which has some people underground. While every fifth employee underground had, on average, a lost time accident in 1976, on surface the average was an accident for "only" every tenth employee.

Mechanization and Injuries

Since operations have always been more mechanized on surface than underground, can it be assumed that increasing mechanization underground will lead to a reduction in accidents and greater safety for miners? Existing evidence would suggest otherwise. Even with mechanization the basic mining process remains the same, and mechanization itself adds some new dangers. It is not possible to establish any clear relationship between mechanization and injuries or deaths primarily because little research has gone into the subject. During the period when mechanization was introduced underground on a large scale there was a decline in fatalities and an increase in injuries. It is clear that mechanization does not remove danger; indeed, it may well have a heightening effect on fatalities in the longer term because it introduces industrial diseases resulting from the fumes and gases of diesel equipment into the mines. Some technological advances have clearly been progressive: wet drilling reduced the amount of dust associated with the older dry drilling methods and thereby reduced the incidence of silicosis; rock bolts have proved safer than the old timbering methods. But safety and health technology has not kept pace with production technology. Often there has been little testing of equipment for health and safety before it is put into production; the machine is tested for productivity, and the criterion of use is profitability.

The following interview with a union official who spends a good deal of time concerned with these matters suggests that mechanization has increased industrial accidents in several ways; for example, back injuries, always a problem in the mine because of heavy work under damp and rough conditions, are now appearing among scooptram operators after only three or four years on the job. There is also a suggestion that technologies that would neutralize some of the health problems associated with mechanization, such as enclosed cabs for scooptram operators, have not been introduced.

What kind of injuries are common in the mines now?

A leg injury, a back injury, an arm injury; you are still getting the injuries from falls of loose on bad ground; you are still getting people crushed between moving objects, between the pillar and the moving train or something like that. Those kinds of injuries really don't change a great deal. One of the problems that we do have came about when the jack-leg drill was introduced about 1953, and it has been one of the major contributors to injuries since then.

The driller now is getting back injuries from this jack-leg. He wouldn't have the same kind of injury as before because you now have a machine that jerks off balance or control. You get the guy where the steel breaks as it's going in and he is holding onto it, and if it jerks he gets a sudden wrench and a jab in the lower back. Those kinds of accidents shouldn't happen. There has been a greater frequency of back accidents or back injuries through the jack-leg as opposed to when they were using the liner machine.

The vehicles now are free-moving diesel vehicles. Does that change the frequency of accidents or the kinds of accidents?

It is rather difficult to say. I don't think there has been any great indication of a higher or greater number of casualties from the use of this big diesel equipment. As a matter of fact, I think before, when tramming was on fixed rails, there were more accidents. But there is a thing now about using these scoops: the fellows who are habitually on the scoop, because of the bouncing, they are beginning to have trouble; but they sort of put up with it because there is good money in the job. The only time you hear some complaint is when the guy has to get out of scoop-tramming because of his back. A guy that's run about three to four years on a scooptram probably has got a bad back, just by virtue of the bouncing and the bashing he is getting from being on that machine.

One thing we are now running into, and I suppose it is going to be more and more noticeable, is the fumes coming off these machines. There are some signs that the ventilating is inadequate. I think there will be a lot of chronic bronchitis, asthma, and anything related to that condition.

Is there any monitoring of the air?

A guy can request a test to be run, but I don't think that is good enough. I think there should be an indication to a man that he runs no hazard at all from poor ventilation. He should know that the area where he is working has sufficient air to make the place safe for the quantity of equipment that is being used. For example, if a man goes down into the mine, let's say to the six- or seven-hundred level, what knowledge does he have that the flow of air is sufficient to supply the needs of the one or two pieces of diesel equipment that are operating there?

In the old days, at least in the coal mines, they used to carry canaries.

They used to carry a little mouse or a bird, and that's a thing maybe they should revert back to now.

Is there any indication of the effects of mechanization on health?

They do have some indications because they have studied the effects in Sweden on someone operating the equipment. The guy in Sweden who operates the same piece of equipment as you have here, he sits in a cab; he's got his own air supply; he is in a properly protected environment from the time he gets on that cab until he gets off and as long as he keeps the cab door shut. I understand there is not much injury. They carry their own face mask and oxygen mask; they have an air environment of their own. So that they have that kind of protection. Now if that is really necessary in Sweden, Canadians are no bloody more removed from those high risk occupations than the Swedes are.

It appears that the only way mechanization can reduce the number of deaths and injuries in the mines is by reducing the number of men working underground. It is conceivable, however, that if the machines were properly designed there would be a reduction of the health hazard. The scooptram provides an excellent example. Technology exists (as in Sweden) to enclose the cabs so that the operator has his own source of air and his head is protected from falling loose. But to install these covers would cost over $10,000 a machine, and Inco's management has decided it is not worth the investment. A conscious decision was taken against providing this form of protection for machine operations. There have been other similar decisions. In the selection of drills by management, for instance, productivity takes precedence over safety. Appearing before a Manitoba inquiry into health and safety in mining, Lorre Ames, vice-president of the Manitoba division of Inco, is reported to have said: "'Muffling drills reduces their efficiency while also reducing noise levels. The bonus system has encouraged some miners to remove the mufflers.'" He also said that "concern for productivity has 'affected the purchase of equipment' without mufflers."[17]

A Sudbury miner whose job was driving a unimog, or diesel buggy which transports men and supplies around mechanized mines, reports:

> *That was a job where the shift boss was always on your back. He wanted to be driven everywhere. You always had a line-up this long, and there were always guys wanting you to get something. It was the kind of job that you could get very angry with very quickly. I didn't mind it too much except that the machines were always unsafe. Like there was always something wrong with the damn things. They were smoking, and I was beginning not to eat my lunch. I felt nauseated. It was just driving that machine; it was hot all the time. You were sitting right by the engine, hot all the time, lots of fumes.*

The most obvious danger from diesel equipment is the fumes. Air must be forced through the mines to clear the fumes, but this costs a great deal of money. Moreover, when one section of the mine has adequate ventilation the dead-end drifts or the working stopes may lack sufficient air to remove fumes.

Less obvious is the danger caused by the machines as they move through the mine. They all work in dark areas and have only their headlights to guide them. One Thompson miner stated that "mechanical stopes are the most dangerous of all. You had to really work in pillar stopes, but you were more aware of what was happening because it's a lot smaller area. But mechanical stopes – they're so goddamn huge that you are not aware of what's happening." The noise and moving lights cause confusion; the environment is completely different from that of the earlier mines where a few people worked together as a crew. In the huge mechanized stopes the operator is never sure who else is there. Accidents in them are likely to be much more serious. The moving equipment will lead to crushings and more cave-ins. As a Steelworker expert in health and safety said, "With heavy equipment you don't get pinched, you get injured. You get seriously injured whenever you get hit by it. So there are no longer accidents the company can hide. If a guy got hit by a scooptram you wouldn't put him in the lunchroom for a couple of days, you'd take him to hospital."

* * * * *

Two Sudbury miners interviewed on mechanization and health:

From the health and safety side, do you think the introduction of diesel equipment has improved conditions?

Miner 1. *No. I think we are going to see real serious problems sometime down the road. People are going to be affected, their lungs or something.*

Miner 2. *We had a case already of a young fellow at Stobie who had been on the scoop for a number of years, and something happened to his lungs. That was from running the scoops, the exhaust.*

Miner 1. *I think it's easier to get hurt by loose with this blast-hole mining because in undercut-and-fill you are always under sand and timber. You now blast and then you have a big opening. You are continually being exposed to loose ground which is a new kind. You have to watch yourself. There may be a big slab there. I don't think it's as safe.*

Miner 2. *I think you are going to see more back injuries in that kind of thing. Really, not because the work is heavier but the vibration from running scooptrams. I think that's got to tear your stomach and your back apart.*

* * * * *

There is an ambiguous relationship between the workers' resistance and the new techniques. The rapid introduction of this technology has coincided with union efforts to improve health and safety conditions. One union leader said, "I think a lot of the technology came about because the union was finally making such strides in the area of health and safety. There is no way that we would put up with those kinds of conditions now. The guys would refuse to work." Mechanization has reduced the amount of physical effort required, and this is clearly progressive. Yet with mechanization have come other, possibly more serious, health problems not anticipated by the workers. As the workers did not control the way technology was introduced, health and safety were not the principal concerns. Instead, the target was production and hence profitability. Managers and workers have different priorities, although some workers can be induced to follow management priorities with the promise of big money through the bonus scheme.

The Politics of Health and Safety

One of the most distressing aspects of the safety and health issue is the way it is used as a political football, by the union as a reason to slow down, by the company as a means of social control. It would go against the company's authority structure to put safety outside the penalty system (see page 265) so that workers could be self-policing. As one veteran miner put it, "Half the injuries are caused by fellows having one eye on their work and the other on the man-way to see if there is a boss or safety engineer or somebody coming along and wondering what the heck would they find this time."

Unions have had to struggle for health and safety regulations. In the early stages of unionization they were primarily concerned with gaining recognition (or fighting for jurisdiction), so that they had few resources either for monitoring or for learning what was needed to improve the miners' safety. Safety standards had to wait for recognition, adequate wages, and decent benefits. Even then there were roadblocks in the way, as a union official recalls:

> *We had to embark on a safety and health program by buying hundreds of dregger meters* [to measure gas emissions] *and the various tubes, training the guys to go in the plants and take readings, have witnesses and document them and take literally hundreds of readings. We had to face the company, having to sneak this equipment in. The company took disciplinary action if they caught anybody and took the equipment off them, and we had to get replacements. We made documentation after documentation with witnesses, and finally, after all the persuasion, the government said yes, there was a problem, and the company had to acknowledge it. Now we have the right to demand readings. But that is the kind of efforts*

we had to get involved in about improving the working environment.

Not until 1969 did the union begin to apply pressure systematically for health and safety regulations, and only in 1975 were major concessions achieved.

* * * * *

A Local 6500 health and safety expert:
Do you think the company volunteers the safety and health concessions?

Oh, no! Up until the last set of negotiations [1975] *the company used safety and health as a bargaining weapon. A company negotiater said last time, when the largest gains were made, "When you look at the health and safety clauses it seems that every letter has been etched out of the granite of management's rights," and he described it exactly. Their kind of attitude is, "It's our prerogative, and we will look after our people. We always have. Just mind your own business." The last time was the first time they really gave anything. The rest was so much hard work. It was a piece of cake last time, safety and health. We got everything in a week – everything we got, anyway.*

Did the Ham Commission hearings have anything to do with this?

Oh, yes! That made all the difference in the world. They were so embarrassed. We got admissions out of the company at the hearings that we have never got before. It was so obvious that the company had fallen down on safety. Not just Inco, but the whole mining industry.

* * * * *

The Ham Commission was an Ontario mine safety inquiry undertaken in 1975. It appeared during the inquiry that there was a considerable gap between the official policy of Inco's senior management and the practice of junior managers. The company's policy was that "no man should work in conditions that are adverse to his health and psyche"; nevertheless, there was constant pressure to produce. Moreover, management stubbornly resists any attempt to loosen its control over the workplace. Prior to the existence of the Ham Commission the union had not had a forum where it could cross-examine company officials. The company held all the prerogatives of power without the accountability.

* * * * *

A Sudbury union official:
Really, the company is not interested in forcing someone to do something which is obviously unsafe. . . . The union, in some cases, has traditionally gone on slowdowns using the safety position. The company is aware of this type of thing, and they are very, very cautious. They want to have the strongest possible weapon in order to offset any kind of movement by the union to have a well-orchestrated type of slowdown using the safety question. . . . If you wanted to work exactly to the company rules, you would never get anything done. You are supposed to scale every little bit of loose, but you could scale up to surface. There is always something coming down, if a guy really wanted to get technical. . . . Unions will use safety effectively in extracting something from the company and use it as a weapon in the same fashion as they are doing to us.

* * * * *

At Sudbury, a very serious danger to the health and safety not only of Inco workers but also of the entire community is the newly constructed Copper Cliff Nickel Refinery. As we have seen, this plant uses a carbonyl gas process for the refining of nickel related to that used in Inco's refinery in Clydach, Wales. Its introduction at Sudbury met little resistance because few people knew of its dangers, especially as used at Sudbury. After the plant was in operation a member of the USWA Local 6500 Health and Safety Committee went to examine the Welsh refinery. What he found was startling. As he says, "You're talking about two completely different setups: a plant that operates at a thousand pounds pressure per square inch and another operating at just atmospheric pressure." Obviously the dangers of leaks are much greater at high pressure. This man described a company-union meeting called over the dangers of carbonyl:

> *I had some stuff* [research] *from the United States where they produced tumours in rats with high doses of nickel carbonyl. I started talking about this kind of thing, and they just watched. They let me finish my little speech, and I said, "These are some of the concerns we have." And the company's doctor said, "Well, I understand that; I have a suggestion to solve the problem." I said, "Well, great. What is it?" He said, "I suggest we keep the rats away from carbonyl." He was serious. And that almost broke the meeting right there. We were ready to leave. This was their head doctor – the guy responsible for health at Inco – and that was his attitude. We used that very well at the interim meetings. They never forgot that for a long time. His attitude was, "You dirty workers, sitting there questioning me. How dare you do that."*

The possibility of a massive escape of the gas into a community that is not prepared to deal with it is a constant risk.

More recent evidence suggests a relationship between carbonyl and cancer. A study of Inco workers in Clydach covering the period 1944–72 found fifty-six cases of nose cancer and 135 cases of lung cancer whereas the expected rates were only 2.3 and 27.4 respectively.[18] The death of a CCNR worker who may have died of carbonyl poisoning has been under investigation. He and two other men were repairing a tank. There was only one mask for the three men, and its air lines did not reach all the way into the tank. One of the men recalls, "About half way through the job I said I had a hard time breathing." The three men were sent home, and one died a few days later. Carbonyl inhalation can cause death in four to eleven days by cutting off oxygen from the lungs. There are antidotes, but the dead man did not receive any.[19] If it is established that he died as a result of carbonyl poisoning it will be the first such case at CCNR and the eighty-third death of an Inco worker in the previous ten years.

* * * * *

Two Sudbury miners, one with ten and the other with thirty years in the mines:

Miner 1. *We fought for years to get a place to wash our hands.*

Miner 2. *That's right. Just in the last ten years we got hot water.*

Miner 1. *So we wouldn't have to wash our hands over the fountain we drink in. When we asked for hot water, they laughed at us.*

Miner 2. *In the last five years we got food warmers and refrigerators.*

Miner 1. *The guys used to tie a can of soup on the light bulb with a piece of wire so it would get warm. We got filters put on the drinking water. We have got a place to wash our hands and soap and towels. They thought we were animals, that's all.*

Miner 2. *Ever since this industry started people have had to eat their lunch on their lap. So you drill for five hours and you got to eat lunch, and you are soaked right through the ass and you try to eat your lunch off your lap – never eating on a table.*

* * * * *

Virtually all the gains made by workers in the mines have been won in a fight with the company. Even an issue such as the wearing of safety glasses can become a struggle depending on the way it is viewed by the workers. In fact, the safety glasses became a symbol of the miners' resistance to company pressure. Appearing before the Ham Commission, Colin Lambert of USWA Local 6500 said, "There was a tremendous struggle to get the men to accept safety glasses and they have pretty well accepted safety glasses now, but where do you draw the line, do you wear your safety glasses from the lunch room to the cage, do you wear them in the lunch room. It's a judgement called on a lot of guys and they figure their judgement is better than other people's."[20]

Why should an apparent safety measure become an issue? Lambert went on to explain:

> In a lot of cases these rules have been used to intimidate a worker who is not in his supervisor's good grace or who is not working as hard as the supervisor expects. This man will receive penalties for the slightest infraction while other workers will break rules with impunity because they are producing.
>
> It is this inconsistency that brings the Company's safety programs into disrepute. The employee does not consider these rules as safety guidelines for his protection. He feels instead that these rules are part of the Company's hold over him.[21]

One recent instance seems to confirm this charge. In the summer of 1978 the chairman of USWA Local 6500's Inquest Committee, Joel Dworski, was fired for "violations of mining procedure." He had been a thorn in the company's side at several inquests and had called for the resignation of a mining inspector. He is alleged to have broken the rules by leaving a bundle of bolts in a box used for transporting material between levels. As a result he was penalized; he is reported to have said, "I was called to the Ministry [of Labour] and grilled about it for five hours on Thursday. Then the next day I reported for work and was told I was dismissed for being an 'unsatisfactory employee.'"[22] The union filed a grievance over the case, but because of the strike it was not heard. In the meantime, Joel Dworski was out of work.

Finally, the Steelworkers charge that safety is used by supervisors to discredit union stewards within the mines. Before the Ham Commission, Colin Lambert argued: "Another subtle way that the supervisors discourage complaints from union officials is that they tell the men on the job that they are being made to comply with the safety rules because a Union official complained. They do not tell the men that they should be following the rules because it is the safe way to work. They insist that the rules be followed only because a Union official complained."[23]

Safety has obviously become a weapon in the war between

the management and the union. It is often used as a cover by both parties. Given the existing authority structure between management and workers there is little likelihood of preventing such struggles. Under it the only possibility for safer conditions would come from greater government intervention, and the governments' records on this have been spotty at best.

The Mines Accident Prevention Association of Ontario (MAPAO) is funded by the Workmen's Compensation Board. It is interesting that "closely related to the MAPAO, and indeed sharing the services of the same executive director, is the Ontario Mining Association [OMA], a voluntary industrial association of corporations operating mines, reduction plants, or like businesses. The OMA is the older organization, and the MAPAO was formed on its initiative."[24] MAPAO is the organization the Ontario government has chosen to implement safety. There is also a close relationship between the mining companies and the Ontario Ministry of Natural Resources. For example, "Of 19 district and chief engineers in the engineering branch of the department concerned with mine regulation, in 1972 no less than 16 were former mining company employees."[25] The Ham Commission argued that the legal system for health and safety in mines in Ontario involves a "strong reliance being placed by the government on the self-regulatory initiatives of industry through individual companies and collectively through the educational activities and moral suasion of the Mines Accident Prevention Association."[26] The companies are expected to regulate themselves since MAPAO is simply an arm of the Ontario Mining Association, which is the collective expression of Ontario's mining companies.

The Ministry of Natural Resources and Mines and its Mines Engineering Branch, according to the Ham Commission, have "fulfilled the dual role of inspecting mines and plants in accordance with the provisions of the Mining Act and through consultation establishing technical standards and codes or requirements for the industry. Such consultation between this branch and the industry has come to be seen by the unions as, at best, accommodation of interests and at worst, collusion."[27] Certainly during the hearings there was devastating criticism of the Ministry of Natural Resources and the mines inspectors by Colin Lambert, chairman of the Safety and Health Committee, USWA Local 6500. He provided a well-documented case of bias on the part of the state in favour of the mining companies. He said, under cross-examination: "We feel there is a conflict of interest within your Ministry simply because the Department of Natural Resources is a department that is set up for the maximization of exploitation of minerals within this province, and if you have a comprehensive safety program you are going to slow up this exploitation, so there is a conflict of interest."[28] The commission report noted that MAPAO provides the ministry with "guidelines" for dust levels and observed that the Mining Act contains no statutory standards. But the commission chose to continue with a hands-off approach: "Self-regulation is preferable to intervention by government. The Commission believes the principle of self-regulation is sound, but the associated issue of accountability of both industry and government must be clear."[29] The issue of accountability was not, however, made clear. What was clear was the committee's recommendation: "That the legal framework for health and safety of workers in mines continue to recognize the importance of a significant component of collective self-regulation by the industry as a whole achieved through a Mines Health and Safety Association."[30] This was too weak even for the Ontario government. They enacted Bill 139 (Employees' Health and Safety Act) as an interim measure to be replaced eventually by an omnibus health and safety act. Bill 139 provided for the right of employees to refuse work for safety reasons, joint union-management safety committees, and placing health and safety under the Ministry of Labour. The actual administration of the act, however, has not used the teeth apparently provided by the legislation.

The following is an interview with a union official involved with health and safety:

> Ham recommended that the company retain its right to determine whether conditions were safe or not, and Bill 139 went half way and said the worker has a right to refuse. Now the debate is over whether the worker is going to be penalized for that and whether he is going to be penalized by being sent home for the day, not being paid, etc., for exercising his right.

Well, other industries are obeying Bill 139, and does that mean that the miner is less responsible than [workers in] *other industries or is the manufacturing company and the construction company more conscientious than the mining industry, because workers have the right?*

Here, I understand, the system is that if the miner refuses to work on the grounds that it is not safe, the company can go to another miner and say, "You do that job," and if he refuses, the company can continue to look until they find someone who does accept that job.

Theoretically that is possible.

Is there any onus on the company to investigate whether the condition is indeed safe or to test?

My understanding is that once the safety hazard has been presented to the foreman, it is his particular business to find out if it is safe. He might declare it safe. The man then can call for a further opinion, and it is supposed to come from the mining engineer. In the meantime, if one person considers it unsafe, it doesn't seem right that another person's life should be put in jeopardy.

It seems that the onus is continually on each individual worker to refuse.

That's right, but that shouldn't be.

Do you think some people would be afraid to refuse to work if they were on penalties or steps?*

There's that kind of – how would you call it? – risk involvement the guy is put through. If you don't do this, you are going to be sent home with a possible Step Four. Does he stay on the shift, or pack up, or does he stay secure? Well, he doesn't know, and is he going to take the risk of finding out?

The legislation does not prevent mining companies from putting workers into areas that other miners have said are unsafe. The union says:

* The penalty system, involving a series of steps imposed by management on workers for infractions of rules, is described on pp. 265-68.

> Unfortunately, our early experience with Bill 139 has been disappointing. The Ministry of Labour has chosen to administer the Bill in what seems to us a closed and negative way. Instead of automatically extending to existing safety committees the rights and powers set out in the Bill, the Ministry of Labour has taken the position that our locals have to negotiate with management for these rights. The same goes for provision of safety representatives in the workplace. Management, on the other hand, has taken the position that although Bill 139 says an employee can refuse work for safety reasons and shall not be "penalized" for it, the company can put the man on another job at lower pay or even send him home without this being called a penalty![31]

It is too soon to analyse the implications of the new bill, but at least in some respects it appears to be regressive. It repeals, for example, a fifty-nine-year-old law that makes a coroner's inquest mandatory for all fatalities that occur in mining.[32] In all parts of Canada, health and safety legislation for mining has been fragmentary and unclear, particularly with respect to toxic substances in the workplace. Health regulations on some subjects differ widely from province to province, and there are no national health and safety standards.

The politics of health and safety exist outside company-union relations as well, particularly with respect to the physical environment. Since the late 1960s there has been increasing pressure on the mining companies to clean up the environmental messes they have been creating. Companies resist because to do so would add to their cost of production. They can only be made to comply through government orders. Inco has been engaged in a number of clean-up projects. Since 1976 about 100 million gallons of the 133 million gallons of water it uses daily are recycled, thus reducing the amount of direct waste released into the water system. A Canadian Industries acid plant, which is the largest in North America, has been constructed adjacent to the Iron Ore Recovery Plant to turn the sulphur gas into 2,700 tons of sulphuric acid a day. According to an Inco official, "environmental restraints on SO_2 emissions forced us to put our roaster sulphur gas into acid. Prior to 1958 we just exhausted all our sulphur gas into the atmosphere. . . . This plant wouldn't be built today. . . . The work force of some six hundred people just wouldn't exist because you couldn't justify it. It was built before the environmental restraints." The most obvious symbol of Inco's "environmental restraint" is the 1,250-foot superstack, the world's largest chimney, built in 1972, which stands above the Copper Cliff Smelter. But the stack has a fundamental shortcoming: it does not reduce pollution. It simply spreads it over a greater distance, from five hundred to a thousand miles.

A federal government study, marked "For Internal Use Only," measured the cost of sulphur dioxide emissions in Sudbury by examining their effects on health, property, and vegetation and concluded that each year $465.9 million in damage is caused. Inco was responsible for 84 per cent of this damage;[33] indeed, Inco is the largest single source of sulphur dioxide emissions in North America. In 1970 it was releasing 5,200 tons of sulphur dioxide *a day* into the air; emissions were reduced to a daily 4,400 tons in 1974 and 3,600 tons in 1978. In large part the reductions were the result of recovery of the sulphuric acid by CIL plants.

Using Inco figures, the government study determined that it would cost $200 million to reduce these emissions to 2,200 tons a day and $500 million to reduce them to 1,600 tons a day. In fact, technology exists to eliminate these emissions almost entirely. The problem, from the company's perspective, is the cost. A useful product, sulphuric acid, can be produced, but as the sulphuric acid market is already flooded there would be no profit in making more. The company's solution has been simply to release the gas into the air.

The Ontario government ordered Inco in 1970 to cut its emissions to 750 tons a day by the end of 1978. Inco did not fully comply, continuing to dump 3,600 tons a day into the atmosphere. In July, 1978, the government decided to rescind its order and licensed Inco to continue at the rate of 3,600 tons a day for another four years with no clear goal for the end of the licence in 1982. Why did the Ontario government take such a position? Appearing before an Ontario legislative committee, Stewart Warner, an Inco vice-president, argued that his company was not responsible for the acid rain (caused by sulphur dioxide) that has killed at least 140 lakes in northern Ontario.

He also told the committee that if Inco had been required to adhere to the order to reduce emissions, the company would have had to close seven of its ten mines and four plants plus parts of another two in Sudbury, reducing its labour force to about 6,500 employees.[34] Alternatively, if the federal study noted above were followed, it could have spent something over $500 million. As it is, the sulphur dioxide is causing nearly that much damage *every year* to other people and their property, but this, of course, is not Inco's cost. It is paid by everyone else.

Safety and the protection of health are inextricably bound to class interests. They are part of the power to define the work process and organization of work. The priorities of management and workers differ fundamentally; managers do not put their lives on the line, but mine workers do. In the workplace safety is used as a weapon in the class struggle by both sides. Outside the workplace management passes on to the community part of the cost of production in the form of pollution and environmental damage. The state in most instances sanctions and legitimizes the right of capital to control the workplace and pollute the environment.

CHAPTER EIGHT

Keeping Labour in Its Place: Managerial Strategies

Changes in the structure and technology of work underground and on surface that have taken place at Inco's various operations were outlined in Chapters 5 and 6. To explain these changes and understand their implications it is necessary to examine more systematically management's various strategies of containment. Managerial strategies for control over the work process are not monolithic at Inco. In part, current strategies are inherited from the past, and middle management often continues to subscribe to particular practices even when senior management no longer formally sanctions them. Some plants or mines that are of recent vintage represent "new" experiments in industrial relations. Moreover, there are important differences of opinion within management along with variations in the workplaces themselves. The settings of Port Colborne, Sudbury, and Thompson each demand different strategies because management in these locations confronts different forms of worker resistance. Underground and surface operations are radically different, as we have seen, and there are many generations of traditions within these various settings.

When confronted with labour unrest, high turnover, and workers' demands, management responds with strategies to minimize these conditions. The strategies of internationalization, diversification of operations, maximizing control over markets and prices, and capitalization or the introduction of technology to replace workers have already been discussed. A fifth strategy is many sided but has a single purpose – the social control of workers.

The "drive for profitability" dictates the actions of capitalists in their attempt to control labour and increase its output and within the workplace has two main expressions. One, as already illustrated, is the introduction of machinery to replace labour and increase its output wherever it is profitable. Another is the application of social means to control labour's behaviour and expand its productivity. Constraints primarily have to do with capital's political relationship with workers and regulations governing the conduct of their work. The latter may be rules and penalty systems governing everything from absenteeism to the workers' behaviour (such as horseplay or respect for superiors). They also include directions issued through a clear chain of command about the tasks the worker is required to perform and the means by which he is to perform them. Usually these constraints are spelled out in contracts between the company and union. The penalties can range from reprimands to sackings, all intended as lessons for other workers. On another level, management's containment policies include accumulation of large stockpiles of products to help them weather strikes and the use of layoffs prior to negotiations to weaken the union. Facilitating strategies, which are not incompatible with those of constraint, are primarily supplying workers with equipment or techniques that will increase labour's productivity, various forms of training, personnel relations practices, and the bonus system. Resistance by workers, on the other hand, is exhibited in high turnover rates in industry, absenteeism, sabotage, work slowdowns, wildcat walkouts, and strikes and will be examined in the next chapter.

The introduction of detail labour as a managerial control strategy is accomplished, according to Harry Braverman, by application of the principles of Taylorism* as a conscious at-

* Taylorism is the term used to denote scientific management formulated by Frederick W. Taylor in which management attempts to control the labour process by job analysis and time study of each element of a task. Its essential characteristic is the separation of skills from workers and transfer of control to management. See Reginald Whitaker, "Scientific Management Theory as Political Ideology," *Studies in Political Economy* 2 (Autumn 1979).

tempt by capitalists to resolve problems of worker control and to squeeze the maximum productivity out of the labour power they have purchased. These principles are "dissociation of the labor process from the skills of the workers . . . separation of concept from execution . . . [and] monopoly over knowledge to control each step of the labor process and its mode of execution."[1] The result is the separation of mental and manual aspects of labour so that work is planned in the administration and executed in the plant. It is not technology *per se* that causes this separation but rather the dominating relation of capital over labour, which seeks to maximize profitability. Of lesser importance, according to Braverman, but with the same purpose, is the use of social engineering focused on the social (as opposed to technical) organization of work and the psychology of workers. This is designed to counteract "difficulties raised by the reactions conscious and unconscious" to the degradation of work.

In the contest between capital and labour, labour resists the domination of capital and represents an important counterpressure. Andrew Friedman stresses that workers have independent wills and are alienated by the work process; thus they organize to resist managerial power. When workers resist this power it is not necessarily the "best strategy" for managers to separate the conception and execution of work, as Braverman suggests. Rather, "if the costs of scientific management in terms of worker resistance or lost flexibility are too great, alternative strategies will be tried" with the organization of work, particularly when technical changes in work occur frequently.[2] Friedman's main contribution is his argument that the degree of worker resistance places constraints on management, which is pressured on the one hand to accumulate capital and on the other to contain labour. Friedman argues: "While top managers are clearly those 'in control' of productive activity [in that they, in their role as representatives of capital, continue to initiate changes in work arrangements and continue to exercise authority over the work activity of others], their freedom to create and recreate this activity is severely limited by the relations of capitalist production."[3] In contrast to Braverman, Friedman argues that "top managers may loosen direct control over work activity as part of a strategy for maintaining or augmenting managerial control over productive activity as a whole (Responsible Autonomy), or they may be forced to loosen direct control as part of a general shift in control over productive activity in favour of the workers."[4] Much of what he is saying is that under particular circumstances it is advantageous to the profitability of capitalists to grant workers some measure of autonomy in carrying out their work, as long as capitalists continue to appropriate the surplus they create.

The argument that Friedman advances is that Direct Control strategies, based on the separation of conception and execution, were developed by capitalists to break down worker resistance based on skill levels, allowing management to appropriate the workers' knowledge and the power that went with it. This lost much of its effectiveness once unionization became prevalent among unskilled or semi-skilled workers. Resistance then shifted from control over skills to the strength of collective organizations. Management was forced to alter its strategy in order to appropriate surplus from the workers. One type of response was the Responsible Autonomy strategy, creating small work-teams to make some decisions about their work. This increases the variety of tasks but also increases work force heterogeneity and stratification into competing groups.

The question to be asked is, how prevalent are these Responsible Autonomy strategies, and do they withstand the test of time? There is little evidence provided by Friedman that they are either widespread or persistent. There is more evidence from the mining industry that managers turn to mechanization or automation rather than accommodate hostile workers. Where innovative and creative workers are required, such as in research and development or in junior management positions, these tactics will be fostered, but most often they are not successful for production, as subsequent evidence will illustrate.

Management will consistently attempt to maximize the output of labour because its goal is profitability. Sometimes, as Braverman suggests, this requires controlling labour and reducing work to detailed specialized labour while at other times, as Friedman suggests, it means capitalizing on the innovative qualities of labour. Which managerial strategy is adopted will depend on the prevailing economic conditions, level of technological development, and political climate between capital and labour. The first strategy is usually adopted for assembly-line operations, for example, while the second is usually adopted for experimental laboratory research. In between there is a great deal of variation, and one task here is to evaluate the relative application of these different strategies in the various work settings in the mining industry and identify the direction of change.

Controlling Labour Force Requirements

Boom-and-bust cycles more dramatic than those in any other major industry affect miners very directly. They are made up of a pattern of oscillation between labour shortage and labour surplus. Sometimes mining companies hire for maximum production to create stockpiles; at other times they coast on these stockpiles and lay miners off. At Inco, superimposed on this general pattern there has been a steady decline in the total number of workers since the late 1960s because of mechanization underground and automation on surface. With machines, fewer workers can put out more production.

Reductions in the size of the labour force can be achieved in one of two ways: by attrition or by layoffs. Attrition means not hiring new recruits to replace those who leave or retire; layoffs mean terminating workers at the time in the company's employ.* Management makes these decisions. Regarding layoffs, they decide the timing, the numbers, and the location of the cutbacks. The only restriction, imposed by the collective bargaining agreements with various union locals, is that the seniority principle apply. (Even here management benefits because the seniority rule means that there is less loss of experience than might be expected.) Otherwise, workers are without protection. Inco management has frequently exercised this power by imposing layoffs just prior to contract

* Laid-off workers have the right of recall, based on seniority, if the company begins to hire again. But in the nickel industry the layoffs tend to be of sufficient duration that workers are forced to seek work elsewhere (unlike the automobile industry where layoffs tend to be shorter and there is provision for paying laid-off workers).

negotiations with unions, so that the layoffs are a weapon in their struggle with workers.

Cutbacks cannot be simply turned on and off. Some control over the number of employees can be gained through attrition, but major layoffs that throw many people out of work require longer-term planning and large-scale schemes, which may disrupt production. The flexibility of the company varies with the prevailing economic conditions. High unemployment makes it easier to rehire a labour force, but when unemployment declines it becomes more difficult. Likewise, when nickel is in demand and there are low reserves the company can ill afford long strikes (as in 1969), but when conditions are the reverse (as in 1978–79), strikes are a relief. Also, long shutdowns or mini-layoffs (as used in the summer of 1978) have less effect on senior workers with longer vacations. Thus management has some restraints on its timing and labour feels the impact of layoffs in varying degrees.

A cyclical pattern with accompanying periods of labour shortage or layoffs has characterized the nickel industry since its inception. Between 1918 and 1921, production of nickel in Sudbury declined from 93 million to 18 million pounds. It climbed back to 110 million pounds in 1929 and fell to 30 million pounds in 1932, only to rise again to 135 million pounds in 1934. Changes in the size of the labour force keep pace with production levels; for example, Inco's work force fell from 7,181 in 1930 to 1,490 in 1932. During the 1958 recession Inco laid off 1,600 men in Sudbury and Port Colborne and reduced the work week by 20 per cent, from a five-day to a four-day week, effectively returning workers' wages to 1951 levels. Following smaller fluctuations, in 1964, 1,850 Sudbury and Port Colborne workers were called back. The work force was reduced by attrition plus 470 layoffs in 1971; in 1972 Inco cut back 4,007 employees and, for the first time, implemented a three-week vacation shutdown in the Ontario division over the summer holidays. Fluctuations in Inco's Sudbury labour force during the seventies are evident in the following figures: the number of hourly paid employees decreased from 20,700 in 1971 to 16,500 in 1974 only to rise again to 18,000 in 1976 and fall, by May 1979, to 13,600.

Through a combination of layoffs and attrition, the Ontario

and Manitoba divisions of Inco were reduced by 4,284 workers in 1978, and summer jobs for students, 1,400 of whom were hired in 1976 and 1,000 in 1977, were abolished. The group let go was larger than Falconbridge Nickel's entire labour force. (Falconbridge itself eliminated 434 jobs in 1977 and announced the layoff of another 750 in March, 1978, out of a labour force of 3,700 workers.) Of the 4,284 Inco workers in Sudbury, 2,800 were hourly and 400 salaried, in Port Colborne 384 were hourly and 50 salaried, in Thompson 550 were hourly and 100 salaried.

These layoffs were not unusual in the history of the nickel industry. They are part of a clear pattern wherein workers are constantly being drawn into and thrown out of the labour force. In Sudbury, 1,261 hourly workers were actually laid off and the other jobs went through retirement and attrition. Of those laid off, over half were single men, and three-quarters were under twenty-five years old. Most of the one hundred women hourly employees were laid off because women tend to have the lowest seniority. Supplementing the layoffs, Inco's Ontario division was shut down for six weeks in the summer of 1978 (there was also a two-week shutdown in Thompson). The Sudbury region lost, through Inco and Falconbridge cutbacks, 4,590 mining jobs from the beginning of 1977; of these, 2,176 were workers actually laid off.

The cutbacks in 1978 hit most severely at Port Colborne because of the low turnover there. Of the 384 hourly workers let go, 97 opted for early retirement, 28 quit, and 250 were laid off, in addition to the fifty salaried workers released. This leaves a total work force of only 890 people in Port Colborne – 750 hourly and 140 salaried – and there were 2,300 hourly workers only a few years ago. In Thompson the layoffs had the least direct effect. The 550 hourly jobs were eliminated completely by attrition (because of the high turnover) but twelve of one hundred salaried employees were laid off and the rest were transferred to hourly jobs (some chose to quit rather than be transferred). Yet the ultimate impact will be devastating because a thousand jobs had already been eliminated in Inco's Thompson operation between 1970 and 1977, and the recent cutbacks come on top of this; moreover, another 660 jobs were eliminated in the city's service sector (including 43 provincial government jobs). The result is an estimated 1,760 fewer jobs and a reduction of the town's population to 17,300 residents.

Related to actual Inco operations, the cutbacks meant that work was suspended at Copper Cliff North, Creighton No. 3, Crean Hill, and Birchtree mines. Work at Levack Mill was also suspended, and in Port Colborne electrolytic nickel production was reduced to production of rounds. Other operations lowered their output. It is important to note that layoffs affect everybody. Some workers are edged out by others of greater seniority moved from closed-down operations. Workers kept on are subject to pay cuts through demotion to lower-classified jobs. An Inco official described how layoffs are managed:

> *It can happen within departments; for instance, if you had to close a specific mine down because it's the lowest-grade mine in your operations, you would close that mine down and you would go into a "force adjustment," wherein you look at all the people in that mine. You may have drillers that have been there for twenty-five years. You may have older maintenance people with a lot of seniority. In a situation like that, you have to look across the board. The older drillers would be relocated to an operating mine, and they could bump people.*
>
> *We have a seniority record and a record of a man's ability and what he has been trained to do and what he has done. We know where he can be used. We can say he has been a driller, he's been a this or that; he's been a maintenance mechanic and a welder in his past history; he's worked here and there. That involves a very long documentation and preparation period. We see one of those every two or three years.*

He also described how "internal adjustments" work through attrition:

> *Then we have adjustments because of changing situations that we call internal force adjustments. For example, an operating department instead of operating five furnaces will go back to four furnaces, or three. What that involves*

is force adjustments within an existing operating department. So, if we have to cut back for a long period of time to align with demand, we would look through all the crews and all the senior people filling the jobs that are left and bumping all the way down, and maybe some of the junior people leaving that department and filling the needs of another department, maybe as labourers.

This man said that Inco avoided layoffs between 1972 and 1977 by taking advantage of attrition and moving workers from job to job in its various operations. "That is definitely company policy," he said. Until late in 1977 this policy prevailed, and Inco accumulated inventories. Top officials then decided to lay off a major portion of the workers.

Layoffs, attrition, and shutdowns all weaken the position of labour. Workers have to use up savings or go into debt when their incomes are reduced because their jobs are classified downward, or when they are without work for up to six weeks, as was the case in Sudbury in the summer of 1978. In this way the company puts financial pressure on the workers with the prospect that they may accept less desirable contracts.

What do prices, markets, and stockpiles have to do with controlling workers? A lot, obviously, if you know how they are related to layoffs, internal force adjustments, and cutbacks. The following are my notes, made July 19, 1977, in Sudbury, well before the 1978 layoffs were announced:

Throughout the company – whether surface or underground, salaried or hourly rated workers – there is the constant knowledge of an eight-month stockpile of nickel. This fact permeates each person's thinking about the future. The world market is slow; Inco has expanded its capacity enormously since the early 1960s: Thompson was opened; many new mines and surface operations at Sudbury have been brought on stream; Indonesia and Guatemala are now producing; sea-bed nodules loom on the horizon. The workers know that cutbacks and layoffs are not only possible but probable. The general decline in turnover at Sudbury has meant the average service of workers there has increased to the point where people with even ten years' service fear for their jobs. Others who are assured of their jobs fear their work groups could be disrupted and they would be forced to move to other operations and "bump" more junior people from their jobs.

These problems are the result of world nickel conditions – particularly the lower consumption of nickel during recessions and peacetime. The reaction to these conditions is not controlled by the workers of Port Colborne, Sudbury, or Thompson, but by senior officers in Toronto and New York. For the union the impact is clearest – [contract] negotiations in 1978 and the day-to-day grievances are weakened. No union, no matter how strong or unified, can sustain a year-long strike necessary to overcome this stockpile. It hangs over the heads of union officers. They are expected to look after the interests of the workers but are without effective means for doing so.

Prices, markets, and stockpiles penetrate right to the face of each stope. The workers know when a push is on or when to go slow. The entire atmosphere of the mines and surface plants depends on these economic factors. Everyone talks about them; people in the mining communities plan their lives around them. The union devises its strategies with them in mind. The company makes its investment decisions – to diversify, to expand, or to mechanize – around them. They are everywhere.

* * * * *

A Sudbury union official, interviewed in the summer of 1977:

Do the workers worry about losing their jobs?

I guess it's in the back of everyone's mind. It's more or less a day-to-day type of thing. You can't plan your life ten years in advance. You never know what's going to happen. It's a very emotionally charged industry. It's not emotions; it's realism and it's pretty hard to understand unless you get involved in it.

* * * * *

Some companies try to control their labour force requirements by contracting work out. Because of strong union pressure Inco has not been very successful in using this tactic in the Ontario division. In Thompson, where there has traditionally been less pressure by the union to protect the availability of jobs, Inco has relied more heavily on contracts with private companies. Some of these are for flux and quarry extraction; disposal of waste rock; trucking of ore from Birchtree Mine in thirty-five-ton trucks owned by Inco but operated and maintained by contractors; care of the giant slag piles behind the Thompson complex; almac raises, a specialized job; sinking of shafts; some diamond drilling; and much of the long-hole drilling. In the smelter some of the maintenance work, particularly welding, is contracted out, as is the making of boxes in the refinery. Such contracts allow the company flexibility and fewer problems in dealing directly with workers and their unions. Generally, however, Inco has not adopted contracting widely. Instead, it has not hesitated to use layoffs and attrition when it feels the necessity to cut back the labour force and massive schemes of recruitment when workers are needed.

The other side of labour force control by cutbacks and contracting out is the problem of recruiting and retraining a labour force. This has been particularly acute in Thompson, where turnover is high, but at various times has also been urgent in Sudbury, particularly in 1968 when the company was hiring thousands of workers, recruiting throughout Canada and even in England through Canada House. The constant issue at Thompson has led to development of recruitment and retraining schemes.

* * * * *

A Thompson personnel officer:
Farmers make beautiful miners because they have the same work patterns. They plant the seed, do their cultivating, and take it off, and they have got to fix their equipment. They are doing the whole of something. If we could only get those kinds of people in sufficient numbers. . . . A guy from down in southern Ontario who's been around the automobile factories or light industry – he's terrible. He really is a lost cause; he is just good for doing that. He's got a narrow piece of the business. But farmers give us another problem because they want to get back to

their crops. We concentrate on Saskatchewan, Manitoba, and to some extent Alberta.

* * * * *

The Thompson program has many aspects. A referral program gives employees a hundred dollars for recommending a relative or friend for employment and another hundred if that person stays a full year. The idea is that if the employees create their own community they will be more inclined to stay. A program of service premiums encourages workers to stay longer. After two months at work they receive 12½ cents an hour on top of their base pay, after four months 25 cents, and after six months 37½ cents (this program originated as a way around the Anti-Inflation Board rollbacks; see page 321). Thompson's biggest source of recruitment is its three full-time recruiters in the field, working mainly through Canada Manpower Centres. Only about 20 per cent of the recruits are "walk-ins" at the Thompson office. To encourage prospective workers to come to Thompson the company gives assistance for transportation and in town maintains Polarises (residences for single men). To keep workers in Thompson longer, the company has a savings plan in which it will match their savings of five dollars a week for five years, after which they may withdraw the whole amount. If they quit short of five years they receive only their own contribution.

Very few Inco employees in Thompson are native people, even though they are the only indigenous people and many are available because unemployment is extremely high. An Inco personnel man described the "problem":

We had a real Indian problem. Trying to employ Indians – they're really trying hard to do that. In fact, we had a fellow in the personnel department learning Cree. He would go around to all the little villages. We learned a tremendous amount. The chances of people staying on would increase twofold if the band council was involved in the selection. But then we would make beautiful mistakes. Somebody who was very low in the hierarchy of people we got from the band council would get a promotion, and all of a sudden everybody would leave, and we didn't understand why. We had broken their caste. . . . Nobody would tell you. They would just disappear, and then this [personnel] fellow would go back to the band council and say, "What happened? What's going on?" And we finally caught on. But then try to report that to a foreman. And try to get the foreman to start to understand what kind of social pressures are going on. So what we then started to do was, rightly or wrongly – we may have mixed feelings about this – we started to set up workplaces for Indians with Indian supervisors. We tried to put them in other locations. But again, they would leave before coming to us and saying, "Look, I don't want to go back on the drill."

The native population has not been tapped as a significant source of labour, nor, for that matter, have women of any racial background.

Obviously, controlling the size of the labour force is a major problem for management. Although the company attempts to deal with it mainly by reducing the number of workers required to maintain production, it is still necessary to have a work force. Instead of maintaining a steady level of production, management has chosen to produce nickel in cycles. These cycles correspond to market demand for nickel, which generally fluctuates with periods of war, but they also match periods when stockpiles are being built by either the company or government agencies. Whenever possible the company accumulates stockpiles prior to contract negotiations and, by publicizing them, attempts to ensure that any strike against the company will have to be a long one. Such tactics create hostility in industrial relations between the company and workers.

Controlling the Workplace

Inside the workplace mining companies create means to control and discipline the work force with the interrelated objectives of maintaining managerial power and maximizing the workers' output. This is done most obviously through supervision and maintaining a clear line of authority between shift bosses and hourly paid workers. Discipline is enforced by means of the penalty system, through which a worker can be fired after a series of warnings. Other less coercive means are also used. Industrial relations practices and "job enrichment" schemes attempt to reduce conflict between management and workers. Another strategy that combines elements of coercion and inducement is the quota system already discussed. But the most widespread control practice underground has been the bonus system (or "incentive," as the company prefers to call it).

We have already examined some aspects of supervision in various underground and surface operations, but it should also be looked at in its relationship to management's strategy for controlling the workplace. When the character of a job changes from craft quality to industrial quality, the managing of work moves from crews of workers to supervisors. For example, underground an independent mining crew originally relied to a large extent on internal direction, but after large-scale mechanization, co-ordination and direction of the work process has come more directly under the control of management. Supervisors are at the lowest level of this direction and surveillance system. They execute the systems created by management.

Workers have ambivalent feelings about supervisors, who represent the control of capital yet as agents of this control are themselves subject to similar pressures.

* * * * *

A Sudbury miner's introduction to shift bosses:
The level boss came and brought us from one lunch room – there were three or four of us – to another lunch room. We walked in the door, and the level boss said to the shift boss, "This is a new bunch of boys for you." And the shift boss looked at us and said, "We need men, not boys." That was my first introduction to the arrogance of Inco's shift bosses.

A Thompson surface worker:
I feel sorry for shift bosses. They get a lot of flak from both ends. I think the company staff should be unionized. Their job's in jeopardy. They are pushed on safety but

then pressured on production. In the majority of cases they are just looking after their asses. They are getting hassles from the foreman, and it's because he's getting hassles from the superintendent. It's just all down the line. The shift bosses have one of the shakiest positions here, so when they get flak, we get flak. It's a cycle. A shift boss who hasn't said anything to you for two weeks will come down on your head, and you know it's not from him.

* * * * *

Many supervisors themselves have the same feelings of being neither fish, flesh, nor fowl. Once promoted from hourly work, they find it difficult to return if they are laid off. They are responsible for what happens in their areas, but they have little control over decisions; basically, they carry out orders. If management does not like their performance, they can be fired because they lack the union protection of the hourly workers. Often they feel lost in the shuffle, not involved in management's decision making, yet responsible for implementing the decisions. Some supervisors overstep their limited power, but for the most part in mining they leave workers alone as long as they produce or until they feel management pressure.

The nature of supervision has changed a great deal since the mines were unionized. The once arbitrary power of shift bosses has been ameliorated somewhat by the collective bargaining agreement and the presence of union stewards. All supervisors now realize that they may be held accountable for their actions.

Inco operates a system of discipline in all the workplaces under which workers may be penalized for certain types of behaviour. According to the company's industrial relations handbook, "Management's right to discipline employees is part of the Collective Bargaining Agreement [CBA] which says that the company has the 'exclusive right and power' to 'suspend, discharge or otherwise discipline employees for just cause.' "[5] The union worker has the right to protest or "grieve" any penalties he is assessed or other violations of the CBA. At one time the procedure was not as closely regulated as it is today. The following conversation with an old Sudbury miner gives some indication of how the penalty system used to work and how change was effected when labour was in short supply:

I'll tell you, an awful lot of the men they hired in the mines were farm boys and from out west. The farmers knew from the first that they had to work; they were accustomed to work, and there was no argument about it. But they didn't like the penalties that were handed out and the guff they had to take from some of the bosses, especially some of the bosses who were pretty arrogant and vicious. Now the thing that stopped that was the war. During the war fellows got kind of independent, and when the boss started to chew a man out for not much reason, the guy told him where to go and gave him his time, and he quit. He either went into industry or into the armed forces. To stop losing so many men, all of a sudden the top supervision clamped down on the lower supervision and said, "Look, you stop this giving of penalties and chewing out the men, bullying the men, and treat them with respect." So for the rest of the war this was a pretty good place to work.

What kind of penalties did they give them?

Anywhere from three-day layoffs to a two-week layoff. Anything more than that, a serious enough thing, they just fired you and that was all there was to it.

What kind of things could you do that brought on penalties?

Well, if you handled muck with your hands; not having a guard rail around the dangerous area; going out under open ground after a blast before it was scaled; or going to work under open ground at all, except to boom it out; smoking in a timbered area; working around the chutes without a safety belt; or standing still beside a train while it was passing. Most of the rules were quite reasonable. What burned a man up was the arrogance and the bawlings-out they used to have to swallow from the supervision, and if you talked back or you used any vulgar language in talking back, that was good enough for a long week or a two-week layoff and even getting fired.

Labour shortages such as those that caused the company to alter its approach to workers during the war made Inco change its industrial relations philosophy in 1970–71 after it had lost a good deal of production during a long strike when markets were strong. About twenty of the old-school employee relations staff were fired or moved elsewhere, and a new "team" was put in their place to "present a whole new face to the union." A senior member of this team said, "We had had a very formal discipline system which involved time off work as discipline. So we changed our system more to one of correction than punitive discipline." When this man was asked whether a particular action necessarily leads to a certain penalty, he responded: "If a person is a good worker, he's a fellow who basically doesn't give you trouble and gets his work done. There's no doubt about it that if he shows up late or goes AWOL, we will sit down and talk about it. But if a person is a bad actor, it's only human nature that the supervisor might say, 'I'm not going to accept your answers any more, and I'm going to discipline you.' "

The current penalty system is quite a complex procedure, and its application varies between Thompson and the Ontario division. It consists of a series of punishments called steps. For further infractions a worker moves up from one step to another, and if he goes beyond a certain number he is discharged. If after a period of time there is no complaint against the worker, the steps come off his record. Steps One and Two are given out by the supervisor, Step Three by the general foreman, and Step Four (beyond which is discharge) by the superintendent. It takes six months to drop from Step Four to Step Three, three months to go from Step Three to Step Two or from Step Two to Step One, and another three months to clear the last step. Penalties are assessed on two grounds, misconduct or absence without leave (AWOL). There is no immediate relationship between a particular action and a certain step; management can impose a penalty at its own discretion. Unionized workers can grieve a penalty if they feel that it violates the CBA (see page 332), but during their initial sixty-day probation period workers can be fired without having gone through either the step system or the grievance procedure. The company frequently uses this right in Thompson, where at any given time over 10 per cent of the workers have less than sixty days' service.

* * * * *

A Thompson miner-union official, questioned about worker-management relations:

What was the company's attitude toward workers before 1970?

The company's attitude in those early days was a lot different than what it is now. The company took the position, "We are the company. Company supervising's our right, and that's it!" They had a situation that they called the bullpen where they used to take guys into the back office, and the shift boss and the highest superintendents would wring them out.

By "wring them out" do you mean bawl them out?

Bawl them out, yes. Bawl them out, and they had no union person present, of course, so they didn't know what was going on. But the company's position then was a lot different. The attitude has changed. I wouldn't say considerably, but it has changed a little bit. I think basically the difference today is that you can meet the company head on, face to face, and sit down and have a meeting with them. I can lift the phone and call Industrial Relations if I have a problem, and they will be looking into it within half an hour or an hour.

* * * * *

We have already referred to the changes in industrial relations at Inco resulting from the costly 1969 strike. Until the mid seventies the industrial relations division commanded a good deal of power, but since then it has lost ground as the financial and production sections of the company have once again regained their positions.

The "new face" at Inco did not appear simply because

management had had a change of heart. It was a response to the strength workers had demonstrated in the 1969 strike. They demanded both improved working conditions and a fairer grievance procedure. Management granted some of these demands, but at the same time it embarked on further strategies that would cut into the effectiveness of the union.

* * * * *

Two Sudbury miners on post-1970 industrial relations:
Miner 1. *There was no change in the attitude of shift bosses. There was no change in the attitude of managers. The only change was there was a buffer* [an industrial relations man] *between the stewards and the upper management of that particular mine. There was a buffer. There was a guy there to do a con job, a little con job.*

Miner 2. *The guy's been the front-line supervisor probably six to eight years and used to the old system: "Do as you are told." How do you change that? You can bring in all the industrial relations men you like. The thing we found was that from 1970 on, you could approach a supervisor and talk to him, and he had to give you an answer. Not that he wanted to. He was forced to because if you went over his head, I guess, he was in trouble. But if he gave you an answer, he would keep on saying, "Well, the shift next week, I'll nail you with a warning." Yes, they did have to talk to you a* little *more.*

* * * * *

From the standpoint of the union, the industrial relations man now present at every mine and plant takes care of some of the simpler grievances but leaves the tougher ones for the union to handle. He can make minor on-the-spot concessions to workers, such as granting time off or not enforcing a penalty for being late or missing a shift. The union steward has only negative power. He can fight arbitrary decisions by supervisors that violate the CBA (and not everything is covered in the CBA), but he cannot grant favours. As a result, the union's day-to-day service to workers is diminished, and the company appropriates part of the service. Workers are encouraged to take their problems to company representatives rather than to the union, and, in many cases, the company can offer concessions not available to the union.

On the other hand, the effectiveness of industrial relations strategies has been extensively undermined by other events. The image of Inco as the protector cannot be sustained in the face of large-scale layoffs such as those in the summer of 1978. Moreover, the requirement to produce was not relaxed on either the workers or the shift bosses. As one Sudbury miner said, "They kept exactly the same pressures on to produce muck. For all the preaching industrial relations and treating men differently, you still had that big monster hanging over your head that said, 'You produce or else.' That hasn't changed to this day." In late 1978 the Inco vice-president for industrial relations, who led the earlier reform movement, was quoted as saying, "There's an awful lot of bitterness now but it really wasn't evident until about a year ago. Not much of that bitterness is directed towards me personally – not even during negotiations – it's the institution [Inco] they hate."[6]*

Finally, it is interesting to note how "job enrichment" schemes have fared. A company official in Port Colborne gave the following assessment:

> *We do try so-called job enrichment schemes here, but we also have new classifications. We have umpteen dozen classifications ranging all the way from the lowest to the highest, with each guy performing a different function and paid a different job rate. What we have done in the Foundry Additives Plant* [part of the anode or furnace department] *and most recently in the Nickel Rounds Building is to define the six or seven different functions and score each one. These may be rated as level three, four, five, six, seven, or eight, and we say, "Fine, all of you guys are going to learn* all *of the operation," and we come*

* On March 6, 1979, an effigy of William Correll, vice-president for industrial relations of Inco Metals and chief management negotiator, was burned by several hundred striking workers. This burning was a response to a letter sent to all employees by the company in an attempt to by-pass union negotiators and Mr. Correll's demand that a vote be held on the company's latest offer (*Globe and Mail*, 7 March 1979:8).

up with a common wage rate, maybe a level six, and we expect everybody to do that and rotate, be it on a daily basis or on a weekly basis, rather than stand on a line doing the same thing day in, day out. But it's funny how it works. Some shifts will operate the way we want, changing either on a daily or weekly basis. Another group may split the job into two parts, the less demeaning and the more demeaning, sort of the operator on one extreme and the fetch-and-carry on the other. On other shifts a guy will say, "God damn it, I've got the most seniority. I'm not going to weigh, to sweep, or hustle my ass. I'm going to operate." So they establish a pecking order, and everybody is receiving the same rate of pay but one guy is doing a level nine job and getting a level six pay, while another guy is doing a level two janitorial job but getting a level six pay. The senior guy is still king, even though it's a job enrichment scheme. I guess we have to blame supervision for allowing this to happen, but then, put yourself in a foreman's spot. As long as the damn work gets done and done correctly, he really doesn't care who does it.

In the traditional form of mining, many underground jobs were already "enriched" to the extent of including a complete cycle of work and self-direction. Now the cycle is being divided into components that are performed using specialized equipment. Some jobs on surface, such as the anode scrap wash, have involved job rotation but usually because one person cannot work at the hammer all shift. For the worker, "job enrichment" is a euphemism for job expansion; he is asked to do more jobs than he did before, not simply a wider variety of jobs. The programs are management implemented, not organized by the workers themselves. A worker cannot go home when he has completed "his" job; he must stay and often has additional work tacked on if there is time left. There is also some evidence that so-called job enrichment will be management's justification for bringing in a new training program that is likely to expand the number of tasks each worker is expected to perform (see page 287).

There is little indication that worker-management relations

have improved significantly at Inco. The dual forces of profitability and containment of labour continue to dominate. Schemes to improve working conditions and reduce the arbitrary power of shift bosses have relaxed some of the most obvious tensions between workers and management, but they have not been sufficient to overcome the pressure to produce or the bitterness of layoffs. The tendency has been toward more rather than less supervision, particularly as changes take place in the organization of work and the direction of work becomes more centralized.

Fundamental to "successful" industrial relations is the company's ability to inspire the workers with confidence that it is out to serve their best interests, at the same time either dulling their confidence in the union or co-opting union leaders into industrial relations schemes. Measured in these terms, Inco's industrial relations program in the decade of the seventies has had only moderate success. Understandably so. It is not possible to inspire trust in workers when they are being laid off or experiencing adjustments to the labour force. Nor is it possible when pressure for production continues to mount. The critical test came late in 1978 when Sudbury workers supported their local in a lengthy strike against the company. Tension between management and workers rose to a level not reached since the end of the sixties.

Early mining companies had contracts with groups of miners based on their production, often supplemented by a minimum daily rate. Not until 1917 did Inco have a firm wage system, supplemented by production "incentives." During this formative period and thereafter the bonus became and remained an institution integral to mining. It was the pot of gold that attracted many men into dangerous and difficult work; for many it turned out to be unattainable because they were unable to produce enough.

In theory, the bonus is a simple incentive or inducement to reward miners for producing ore or doing development work more quickly. In practice it has more complex aspects. To many miners it is a proud symbol setting them apart from most workers in other industries and a measure of their skill and dedication in their craft; it is a justification for taking the risk of working underground. To the unions it is a source of danger, luring miners to work carelessly, and it takes jobs away from other miners. To the companies it is a means of social control, the carrot that reduces the amount of prodding needed in the workplace. It is, in the minds of many, what makes the miners go. What will be the fate of the bonus? Will it be affected by mechanization? These are complex questions, and the answers are currently being debated within both companies and unions, and between them, in the mining industry.

The old contract system gave workers a great deal of control over their work process. It functioned much like a piece-work system in that workers under contract were encouraged to increase production but differed from piece-work because miners' incomes tended to be higher than those of most workers. When employers lowered the rate of pay, workers responded by collectively cutting production and unionizing. The employers then substituted for the contract a bonus system that offered incentives above a set hourly rate. They could directly control the bonus rate, raising or lowering it as they saw fit. Because it was outside the negotiated wage rate, unions could not influence it. Moreover, it was used only in situations where the workers could affect output directly. In early years it was used in some surface operations demanding a great deal of manual labour plus skill or in some critical operations that determined the overall rate of production. Many of these surface jobs have been eliminated and only a few critical-task jobs remain on bonus, mechanization or automation having absorbed the others. There are no surface jobs on bonus in either Thompson or Port Colborne, and in Sudbury only a few jobs in the oldest operations such as the smelter and copper refinery are still on bonus.

About half of those who work underground are on bonus. Underground the miner still exercises a great deal of control, and the company relies on his ability as well as his hard work. The bonus system is found in all underground operations at Inco, but there are important variations resulting from mechanization. The most notable is at Levack West, where the entire mine is on a single bonus; as will be recalled, it is the most mechanized mine. Other mines have had some tasks mechanized, and the bonus system has been adjusted accordingly.

The union can "protect" its members only for wage work. The company has retained control over the bonus, contract, or incentive system, as it is variously called, although workers or the union can grieve the regulation of the system. The workers say that it is best described as an invisible foreman. Given the great interspace between work crews and management, this system forms a bridge. But as tasks are subdivided and mechanization strengthens the co-ordinating role of management, there are other answers. Supervisors have more mobility in the ramp mines and can oversee the workplace with greater ease. And as workers have less control over their rate of production, some managers conclude that the bonus is no longer necessary.

The bonus system normally is based on time worked. Time, in turn, is translated into the miner's hourly rate. If two people have ten hours' bonus due them and one has an hourly rate of four dollars and the other a rate of six dollars, the former will receive forty dollars and the latter sixty dollars for the ten hours. All bonus jobs in a mine are assigned a time limit by the engineering division. For instance, a man may be required to set up a certain number of posts at half an hour each or to drill so many holes in a specified number of hours. The shift boss keeps a record of each task, and at the end of the month an engineering crew comes and tallies the work done. The amount of ore removed is calculated by the number of feet the stope has been driven. Each task in the operation is compared with the actual amount of time worked. The amounts are calculated for the entire crew, but the amount of bonus varies for individuals. If, for example, your output equals 200 hours' worth of work in the month and you actually worked 160 hours, you are entitled to 40 hours' bonus. This would be calculated as 40/160 times your hourly rate or, in this case, a bonus of 25 per cent. For different aspects of the work the method of assessment varies. On trams, for example, the bonus is based on the number of tons hauled from one point to another. In long-hole drilling the rate is determined by the number of feet drilled. All of the systems have the same goal: the greater the productivity, the greater the bonus.

An old Sudbury miner explained the difference between the bonus as it used to be and as it is now:

> *At one time the bonus was calculated as so much per shift in a stope. So every man who spent one shift working the stope would get one shift of bonus, and the new shoveller got the same bonus for a day's work as the stope boss with ten, fifteen, or twenty years' experience, who was pretty well running the stope. We had, at that time, what is known as a guaranteed minimum contract incentive system. I was responsible, in 1966, for suggesting a variation of the standard man-hours system. In the contract system, all the rates were in dollars and cents. In the standard man-hours system, all the rates are in time, so if your crew spends five hundred hours working in the stope during the month and if they calculate you have done seven hundred hours' worth of work, that two hundred hours counts as bonus. It is calculated on your efficiency; not quite that simple, but pretty simple. Mining is different from any other industry. Individual skill, knowledge, experience, and hard work could make a tremendous difference in your efficiency. While one crew may be working along and working themselves to death and making a little bit of bonus, another crew will get twice or three times the work done.*

The relationship between bonus and the skill of the traditional miner is obvious. Hard work does not necessarily lead to more money; technique has a lot to do with whether the miner will end the month with no bonus or five hundred dollars. At least that was true in traditional mining. There is currently a dispute within Inco over whether or not bonus is anachronistic in mechanized mines. In the union, on the other hand, the relationship of the bonus to health and safety is the question. It should be stressed that the bonus system is controlled by the company. The union has nothing directly to do with it; it is not part of their collective agreement nor is it a negotiable matter.

The following interviews contain views of the bonus system expressed by two senior Inco officials, the first an engineer and the second a manager whose main experience has been with personnel. The different emphasis each places on control over production by the miners should be noted.

Interview with an Inco engineer:
A lot of production is on an incentive basis. Most production is done on what you would call a bonus system. In other words, you get a certain pay and there is an incentive; the more you produce, the more you get. That basically is the keystone of all mining in this country.

Is the incentive still as important with the introduction of mechanized equipment?

Levack West is our newest mine and the most modern mine. We think, presently, that it is the most modern mine in the world. It is the only one that the Russians said, when they came last year, was better than anything that they had. I don't know if that is a stamp of approval, but it is the most mechanized one. The whole group is on one incentive contract. Incentive makes that place go. They are kind of self-policing to a degree where if somebody is dogging it over there, well, there is no way that they are going to let him, because it is costing them money. It is the most mechanized mine and there is no way that the incentive doesn't play a very, very strong part in the operation. You plan and make sure that you utilize the time that you are there in a productive manner; the overall planning certainly is directed and supervised, but the individual outlook of the particular worker – nobody is looking over your shoulder, literally, no matter what you are doing, because there is nobody there to do it, though that is true of mining anywhere.

We are doing more work on the incentive system than we used to. Much more. . . . The Swedes have tried an experiment and went off it, and I think the productivity went down. They admit to about 20 per cent, but we think that it went down more. They put everybody on salary and took away the incentive system. They took a big nosedive.

There is something quite different about a mining industry and a miner. He is somebody that works unsupervised. There are not many industries – I don't know of any industry – where you can say that is the way it happens. Underground the normal supervision amounts to maybe, at the most, two visits, possibly ten minutes a day, and the rest of the time he is by himself. He makes his own decisions, he sets his own pattern of work effort, but nobody is looking over his shoulder; nobody comes back at the end of the day to find out how much he did because there is no way to do it physically. The supervisor has an area to cover, and he is fortunate if he manages to visit them twice a day. So the miner has quite a different attitude; he is rather proud of what he does and how he does it.

Interview with an Inco personnel man:
As we went through the mines and talked to the miners, there was a tremendous pride in mining and high skill levels – more so in the shaft mines than in the open pit or in ramp mines, which have more equipment. Mining takes an awful lot of experience. It takes a lot of on-the-job training to become a good miner.

Well, I don't think it is the highest level of skill. I think the pride in the mining job comes from being able to do a complete cycle of work. I think there is a tremendous amount in that. The miner does the whole thing right from the layout through to the drilling, the blasting, the screening. He does the whole cycle and he can see it. He can see the results of what he has done. But drilling isn't hard; we can teach you in two days. Blasting is safety instruction. It's not really that high a skill. And mucking up, that really isn't a very high skill. The component parts, the cycle of work, these are really not high-skill jobs. I think where I am disagreeing with you is that I don't believe that it is really that high a skilled job. It's the pride of the miner that comes from the relaxed supervision. You do the complete cycle of work. I don't think it's complex.

But just being able to drill is not the point; there are some people who can drill five times as fast as someone else. And the key motivation, we've been told, is the bonus in the mines. The bonus keeps people moving.

I don't buy it. I don't buy that at all.

I can see that bonus won't work as well in Thompson (except as an attraction) because I can learn to drill or screen or muck or blast, but I'd have to be there a long time to get a 30 or 40 or 50 per cent bonus out of it.

I firmly believe that it's the work [that keeps miners going]. *They're there by themselves and only see a shift boss once or twice a day. They do the full cycle of a job. And the pride of the mine crew, and the fact that they can make some extra money through a bonus, that's part of it. I don't exclude it completely, but I think they can produce every bit as well, possibly better, without the bonus.*

What happens in a case where at one time a man went through the whole cycle in a stope mine, and now he is doing nothing but in-the-hole drilling all day? What happens if you take the bonus away on that?

We've spent a hundred million dollars on equipment in Copper Cliff mines, and we have a lower productivity. That's in-the-hole drills, raise borers and scooptrams, and everything else. We have lower productivity than we had before in tons per man. Your case of in-the-hole drill – that's one that I don't know whether they have resolved or not yet – that's a fascinating one in that the mine bonus system is based on physical effort, but the in-the-hole drill is 70 per cent machine. It's very much a machine type of operation, and those fellows learned very quickly that once they did seventy-nine feet or something like that, they'd come to the maximum output in their bonus. We know the machine is capable of doing four hundred feet a shift, but they were only doing eighty or somewhere under a hundred. They were trying to change the incentive system to say, "Okay, we will pay you so much per foot," but then you are shifting and you are paying for the machine time and the guy watching. That has a major impact on the total mine incentive system. I think that the mine incentive system in Copper Cliff was okay; it's just out of date. And we have not been able to have it opened too much to scrutiny. Now it pays those people well, and it pays them fairly; but it's not going to stand up to scrutiny.

The irony of this argument is that the only people I have heard agree with you are union executives who would argue similarly, except they say bonus is unsafe; it makes people do things that they shouldn't be doing.

My rationale is different. Mine is people doing the cycle of work; theirs is the safety.

The argument against you would be that you can't say it's 70 per cent machine, 30 per cent man. It's 100 per cent machine and 100 per cent man, because if that man doesn't tend that machine – keep the filings away as they're coming out, keep it clean – that machine's going to break down, and he won't give a damn if it breaks down because he's not on the incentive. But if he is on incentive, he is going to do everything he can to keep that machine operating at optimum.

Well, that says that everybody should be on the incentive system. But we were able to get that guy away from that machine for two and three hours at a stretch. He would just have to go back and check it once in a while.

One case that would support what you are saying is Levack West's ramp mining system. They have moved on to an incentive system for the entire mine. Not individually, but the entire mine, and that looks suspiciously like the beginning of the end of the bonus system. It is very high now, about $8,000 for all the people on bonus underground.

Everybody is on the incentive – the guy cleaning the snow outside, the guy getting the timbers outside. Everybody is on it.

But that looks like the beginning of the end, at least an experiment in that direction; would that begin to signal the end of the bonus system?

That's why they haven't gone any further. We tried very hard to say, "Okay, try another mine. You've got large working areas now, take that working area and make that into a group." I make my statements about mine incentives quite openly.

Conclusions concerning the future of the bonus system obviously cut across the usual worker-management line, although the reasons for arriving at these conclusions differ. The contention of nearly everyone is that many workers are in the mines to make bigger money than they can make elsewhere. Older managers would tend to agree, although they use different language, arguing that as direct supervision is not possible in the mines an incentive is used in its place, and this keeps the miners producing. Some other managers, usually the younger ones with less mining experience but broader industrial relations training, believe that bonus is not money well spent. They think men will produce whether they are on bonus or not, either because the work is intrinsically interesting or because under mechanization the machine sets the pace, so that mine workers can be supervised like employees in most industrial settings. Union officials argue - although not very loudly - that the bonus promotes sloppy work. More important, when people work too fast, cutting corners, they endanger their lives. Both union leaders and workers tend to agree that bonus is "the boss."

Union leaders also reason that the bonus system is costing some miners their jobs. This argument emerged with the layoffs in 1978. Dave Patterson, president of Local 6500 in Sudbury, is quoted as saying, "If our guys wouldn't work on bonus anymore, they could save hundreds of jobs. Laid off miners would be returning to work a lot sooner."[7] The company would appear to agree with the premise of this argument but reacts differently. The bonus system, by increasing each worker's output, reduces the overall cost of production. It is more economical to keep down the numbers of production workers even if they are paid high incomes because there are fewer workers for whom to pay benefits, and the bonus ensures that the workers will perform up to minimum standards. Without the bonus the company would have to spend more money on supervision to maintain production.

Another aspect of the bonus disliked by union officials, particularly in Thompson, is called free work: a shift boss gives a crew of miners some "new hires" to do all the bull-work and nipping in the stope. The stope crew is thus able to produce a lot of ore and receives a large bonus credit, but, union officials complain, this shows favouritism and is not fair to the labourers. The stope crew that gets out a lot of ore is used by stope bosses as an example to other miners. According to one Thompson miner, "These guys – who are supposed to be an example for the rest – they give them all the help possible. So at the end of the month these guys make 50 or 60 per cent bonus and all the other guys make 5 or 10 per cent. They hold out the carrot."

* * * * *

A Thompson miner and union official, asked whether he would like to see the bonus done away with:
Well now, you are trying to put me in the hot seat. If you suggest to miners that you are going to do away with the bonus, they are going to get all upset and uptight and all the rest of it. However, yes, I would like to see the bonus done away with, but I would want to see the wages raised, substantially, in compensation for the bonus.

* * * * *

Miners argue that bonus was greater in traditional manual mining than it is in mechanized mining. When the rates were revised they were set higher so that the maximum bonus in mechanized mining is about 35 per cent. The notable exception is Levack West, the most highly mechanized mine, where everyone is on the same contract and the bonus is about 50 per cent. Union officials believe this operation is being subsidized by the company. In all the other mechanized operations, young miners tend to run the equipment, but the machinery in Levack West is operated by older miners. This mature work force was recruited primarily from Levack Mine, and the miners were given the opportunity to bid to the new mine, thus attracting those with high seniority and experience. Union officials are led to think that the company is using Levack West as an experiment for a new bonus arrangement, and some company officials reinforce the impression.

Levack West may well be indicating the direction of the future. Already the open-pit mines, like Pipe Mine, are off bonus; and they are obviously the most mechanized form of mining. Inco officials have hinted that in place of the present bonus system another form of incentive, such as profit sharing, which would include both surface and underground workers, may be implemented. The likelihood of this occurring depends upon how rapidly further mechanization takes place and how much pressure the unions apply regarding health and safety. If the conditions are right, the management group that would like to see an end to the bonus may win the day. For union officers the risk in openly opposing the bonus is their prospect for re-election. Unless they could offer at the same time substantially higher hourly rates as compensation they are unlikely to receive membership support. At the moment other issues have taken centre stage in union-management negotiations, but when pressure is once again on for production the bonus is likely to become a significant issue.

Controlling Training and Skill Levels

Virtually every hourly rated worker at Inco spends his first days with a broom in his hand on surface or at the end of a shovel underground. After an initial indoctrination he normally joins a labour pool used at the shift boss's discretion. From there he will be transferred to a specific job or he will bid on some training that will make him eligible to bid on jobs as they are posted. At the most basic level a miner receives physical training. New hires are worked hard, the idea being that this will give the shift boss a chance to weed out those physically unable to stand up to the job. Next follows what is essentially defensive training, or learning the dangers of the job. In the past this was taught primarily by other miners, and Inco has only recently incorporated it into its induction. The job skills have also traditionally been passed on by the miners themselves with little of the responsibility falling directly on the company. In recent years the company has had a growing role in training, motivated in part by health and safety pressures but also by the new methods needed for mechanized mining.

C. Hews, appearing before the Ham Commission on behalf of Inco, was questioned on training.

> Q: In the one-week training period may we assume that the new hire is not actually on the production force, he is simply learning during that period of a week?
> A: That is right and that one week is not necessarily that period. If it is required and if he feels uncomfortable or there is some reason to repeat a portion of that then that is done.
> Q: Now then, he takes the test and is successful. The brief yesterday indicated that at that point he is qualified to bid for a driller's job. Would you care to comment. . .
> A: That is correct. At the point in time that he is qualified as a driller, then he is permitted to, in fact, bid, or I should say his application is then accepted on a driller's posting.
> Q: Then is it possible under this system for a man to become a driller on the 8th day after you hire him if he had no previous experience?
> A: Yes it is.[8]

The previous day Colin Lambert of USWA Local 6500 had presented a brief condemning Inco's training system. He said:

> Inco's training of new men is totally inadequate. A man comes to work for Inco as a miner. He is put in a completely foreign environment doing work that he could not possibly have been trained to do before and even the language that pertains to his work is completely foreign to him and he is given one week with an instructor to learn a new trade. After this one week, and quite often this is not a full week as he might be used for other jobs that need doing, he is given a test. This is a written test as to the man's knowledge of certain safety rules. When he completes this test, he is then qualified by Inco as a driller. I personally never have heard of anyone failing this test and going back through the course again. . . . If he gets a driller's job, he will possibly be put with an experienced driller. This driller will be working under a bonus system. He will not have the time to take the new man and train him, in fact he will probably resent the

fact that this is an inexperienced man and be reluctant to even attempt training him as it is going to cost the miner money out of his bonus. The new man has to try and pick up the trade on his own and if he cannot pull his weight with an experienced man, he is usually pulled out of that drift and put elsewhere, probably with another new man.[9]

A survey conducted by Energy, Mines and Resources Canada underscores the limited amount of formal training in the mining industry. It found that "34 per cent of trainees in skills spent less than 20 training hours [fewer than three working days] and only 16 per cent of trainees were enrolled for more than 200 hours [twenty-five days] . . . safety training registers the greatest proportion of trainees. Training is, in general, of short duration with 50 per cent of all trainees (excluding safety training) spending less than one week on training."[10]

The use of existing mining crews to teach novices the necessary skills – because they certainly do not learn them in one week's class – has a disruptive effect on the crews themselves. A Thompson miner commented: "We've got to train these guys. When a new guy gets in the stope you have to take your time to train him. It costs us money. You train somebody and they up and leave." Not only is their bonus affected, but people are seldom allowed to work together for long periods of time. As a Sudbury miner said: "When a couple of fellows got together and they became pretty well experienced and good miners, then one would be the leader and the other the driller. Afterward the bosses would get the other fellow [the driller] to either bid on a leader's job or they would appoint him to a leader's job anyway, so they could put a couple of new men with these two fellows. As a rule, if two fellows worked together for two or three years, well, that was long."

This is an Inco miner's account of the type of training he received:

We reached the school stope, maybe 600 feet underground, and it was unlike any classroom we had been in before. There were two instructors, both older miners, and what they taught us would pay off a lot sooner than any of us realized.

Each day they showed us as much as they possibly could.

But each day, without fail, our instruction was interrupted as we were called to do some shovelling some place, or some pailing cement somewhere else. Some of us never made it underground at all: they were sent somewhere else to work for the day.

We spent about two weeks at school. . . . We were given the test questions in school stope, and we went over them together getting the right answers down on the paper. On Thursday afternoon, we were told to study the test paper overnight to get ready for the examination the next day.

The next day, each of us was called in to the safety engineer's office and given the test orally. It hardly mattered that many of us spoke only French, or Czech, or Polish – the test was a farce anyway. We were asked questions of no significance, we were in and out in a couple of minutes. Now we were qualified drillers and air-slushermen. We had graduated.[11]

At Thompson, miners consistently reported that they received very little training before they entered the mines. The following account illustrates a miner's induction:

At that time [1968] *they kept you on surface for half a day and they gave you your gear, your hard hat, your boots, and they explained to you about certain procedures of reporting injuries, for instance, what it was like underground, the dos and don'ts of the job. They gave you a little booklet on it. Basically, that was it. . . . I started off underground as a shoveller in the ditch. Everybody starts the same way. Underground jargon for that is a "jigaboo." You sand out the ditch so the water can run, and periodically you are taken out of the ditch. You may be on the shovel, or the banjo as it is called, for a considerable period of time, three weeks or a month. Then you are taken off and start to do other jobs like nipping material into the stope, or you may be working the stope doing various jobs, giving the experienced people a hand.*

After the initial breaking-in period, neophyte miners are sent to a stope school; as another Thompson miner said, "There you didn't learn too much because you don't learn mining in two weeks; mining takes a few years to learn the job properly." This miner said that after six weeks of shovelling and his time in a stope school, which was interspersed with more shovelling, he was put in a stope with one other miner; "he was showing me the ropes and then he bid off, and I was appointed stope leader after three months. I had a basic idea, but as far as skill goes I was a bull in a china shop. It was all work and nothing achieved. I learned the hard way. I watched the cross shift to see what they did."

Since that time there have been changes in Thompson's training program. Three deaths underground in the first three months of 1977 sparked a more concerted training effort, and a five-day training program covering the basics of mining was instituted. The recruit is taken on a tour of the mine the first day, and "the rest of that first week, the new employees work in the mine. The new employees return to the classroom for one day in each of the following five weeks. Scaling, drilling, timbering, machine operating, sandfill operations and other mining procedures are explained and demonstrated."[12] This system of instruction one day a week for five weeks is obviously related to Thompson's high turnover. If they went to school for a week straight, many new hires would never even see a production stope before they quit. In Sudbury the program has also been extended to include two weeks' initial indoctrination and then at some later point a month's training in drilling. Much of the pressure for greater initial training has come from health and safety inquiries.

As mining is an industry with high turnover rates, the cost of labour is increased considerably by the price of training, estimated at about a thousand dollars per person. Recently two types of mining training have appeared as reactions to the increasing mechanization of mines and the introduction of process technology on surface, reactions that differ radically from one another. Inco is rapidly moving into a modular training system as an expression of managerial strategy. The United Steelworkers are pushing for an apprenticeship system called miner-as-a-trade as an expression of workers' resistance (see page 341). These developments appear to be incompatible, each striving to take training in a different direction.

In 1970 Inco undertook a pilot experiment in "functional

modular training." Called the Instrumentation Training Program, it instructed forty instrument mechanics in the maintenance of instrument control systems for use in all its automated plants. First all processes and instrument equipment had to be surveyed and modules designed to teach the mechanics how to test and repair the equipment. The fact that this was necessary reflects the increased use of instrumentation in Inco surface operations. Traditionally the training for this work was a four-year apprenticeship, but when new techniques were introduced at the Copper Cliff Nickel Refinery and elsewhere, a shorter training period had to be devised. According to a senior Inco official involved in implementing the system, "the increased needs were imposed by new technology; traditional training couldn't respond." The use of the modular system shortened the qualification time from four to two years, obviously an advantage to the company. This modular training program is a registered trade in Ontario, but it is a non-regulated trade, which means that the government does not supervise the course content. This is different from the apprenticeship program, where the government specifies the content and provides a broader training package. Here was the basis for Inco's objection to apprenticeships: they contained much training not needed for specific work at Inco plants. Management wanted something tailor-made.

In 1972 Inco again used a modular training program, this time for the processes at Copper Cliff Nickel Refinery and the electrowinning plant at Copper Cliff Copper Refinery. As has been seen, modular training was not entirely successful there because operators found themselves being required to do maintenance work at operators' rates. Inco officials feel the problem does not lie with the modular training: "All the training will do is get you to do a job. If you say that it includes the maintenance of the equipment you are operating, well, fine, but it doesn't have to. It was a management decision that was taken to incorporate those kinds of things [like maintenance] in the job, and that did result." It is true, however, that the modular training system gives management opportunities to push operators into more maintenance work and to expand the tasks contained in each job.

Inco has also used modular training in a supervisor induction program and a mines induction and drilling program. In March, 1977, it was announced that its use would be extended across the entire Ontario division. In setting up the system Inco called on a U.S.-based firm, Management Training Systems (in Canada called Tectonics), to survey the operations and assist in the preparation and use of MTS Training Manuals. These same consultants had worked earlier in Canada for Polysar in the petro-chemical industry.

Modular training is a form of "Gestalt structured functional training," according to the report the consultants prepared for Inco.[13] The theory requires every task to be broken down into its parts; these parts then become interchangeable and can be arranged in a variety of ways. "Functional training is training based on the physical plant; the basis for determining what kind of training is needed is the basic process and equipment," according to an Inco person responsible for the program. Every process and piece of equipment must be documented in a systematic way and inventoried. Each operation is viewed as a process based on a series of pieces of equipment. In the language used earlier, the purpose of the training is to make a science out of an art by removing the mystique of the task. Production is rationalized.

The central principle of MTS is that "*the job defines the training and not vice versa.*"[14] A manual is produced for each piece of equipment that can be applied to equipment of the same kind in any operation and includes information on how to operate and maintain the equipment. It is administered largely by self-learning. The MTS report foresees that "many operators will learn (or [be] asked to learn) to do things that do not fall within their present duties."[15] Maintenance workers will use many more manuals than production workers, but production workers will require more than one manual for training. The training unit is the equipment, not the person. In a trade there is a common core of skills, such as learning the selection and use of tools. Principles and techniques are learned and then adapted to the situation. The *training* is broad. In modular training the situation is determined first; the training is more immediate and "practical." The manuals are, according to an Inco official, "filled with tasks, but the training is not split off into small tasks but into manufacturing steps." There will be a common way for performing each task, operation, and process.

Under the collective bargaining agreement management reserves what are called management rights. The CBA between Inco and USWA Local 6166 in Thompson, for example, reads: "It shall be the sole and exclusive prerogative of the Company to decide on the location of its plants and mines, the products to be mined and/or processed, the schedules of production, the methods of mining and processing used, the number of employees needed by the Company at any time, operating techniques, methods, machinery and equipment, and to exercise jurisdiction over all operations, buildings, machinery and tools."[16] Therefore, under MTS it is management's responsibility to establish jobs, that is to put the manuals together, and "management decides"; the matter is not negotiable, as far as Inco officials are concerned.*

From March, 1977, Inco's personnel department had an eighteen-month mandate to prepare manuals for all the plants and mines in the Ontario division. There were thirty writers working full time, hundreds of people feeding information to the writers, and five MTS engineers working mainly on the format of the manuals and assisting with the actual training. About 40 per cent of the manuals are "off-the-shelf" books already developed by MTS for other companies, but nearly 60 per cent had to be custom made. The company expects to produce about a thousand manuals in all, about two hundred of which are for the mines. A costly implementation program is in prospect, but the key gain for Inco will be flexibility in using its labour force.

Currently Inco has 256 job classifications in Sudbury. One of the aims of MTS is to reduce the number of categories to ten or less.[17] The intention is to build the training into the actual operations. The modules are not time based; they are performance based. There are many problems with the system, but one of the more important ones in mining is the difference between knowing how to do something and how long it actually takes to do it. It is one thing to do something and quite another

* MTS does, however, dovetail with the Co-operative Wage System (CWS) to be discussed shortly, *but the determination of the job is still the responsibility of management.*

to do it quickly and under the variety of conditions found underground. This is particularly true of the more traditional tasks in mining where there are different rates of speed between even experienced miners. The *rate* of performance is not necessarily built into the MTS package, but with modular training it is possible to set specific standards. It is also possible to add maintenance to the operator's job or leave it out, but the system invites management to expand an operator's tasks just as it invites setting rates and raising standards of performance because management will now know exactly how long it takes to do a specific operation. There is also a great temptation for the company to fill in free time with additional tasks it knows the workers can do and establish quotas for output.

Some union officials are aware of the implications of the MTS system but feel powerless to do much about it, given the constraints of the CBA. A union official indicated what he foresaw:

> *I believe that Inco is really trying to camouflage what they are trying to do by saying, "We want to train our people right." Well, they have a group of people drafting these training manuals. The end result of these training modules is to have each man able to do three or more jobs. To me these training modules have a long-range effect to them. It's going to be a quota system. Some place along the line the company says: "You start here; it takes you four hours to get your quota out here. Once you are done with that we put you over here, where we know it takes four hours for quota." Now they've really got you utilized for eight hours. To me that's what these modules are all about. There is a continuous problem of operations people doing maintenance. That's what is called job enrichment. It's really job expansion; the company is trying to get more production out of individuals.*

Modular training in a high-turnover industry like mining makes it easy for the company to replace trained people. When skilled workers are lost through layoffs, long strikes, and turnovers, Inco can rapidly replace them with MTS trainees. Although this was not the purpose of the most recent layoffs and strike, it could be a secondary benefit for the company. This system should decrease start-up problems considerably, even if the company loses experienced workers during the disruptions.

It is clear why MTS is coming into use now. A senior Inco official said, "It's a reaction to technology, no question. We've got more training to do on different kinds of things, and unless we get a good system established we'll be scratching our heads. It's a reaction to the changing technology of training as well." Like all technology, MTS has a high initial cost, but "in the long run it should be cheaper to pay for training." MTS is a reaction to mining technology, but its application is practical only for highly mechanized and automated tasks. One can see how such a training system will work for operating a raise borer; it would not work as readily for a raise crew, but then again, there will soon be no raise crews left.

The MTS approach runs contrary to traditional apprenticeships and on-the-job experience. MTS people know this and present it as one of their strengths. In their report to Inco they argue:

> Inco's apprentice programs are typical in that they don't produce people who can do the work that needs to be done and they take way too long to do that. There appears to be a power play about as follows:
> 1. Set up apprentice programs.
> 2. Obtain "certification."
> 3. Legislate requirements to allow only "certified" people do the work.
>
> Unfortunately none of this has much to do with producing metals faster, safer, cheaper, without damage to the environment. Apprentice programs have non-economic justification, if any.[18]

By implication, of course MTS can solve all these problems ("damage to the environment"!). But it too is involved in a power play, the selling of itself and its service to Inco. As yet there has been no large-scale showdown with the union over this threat to apprenticeship (although a skirmish occurred in the Copper Cliff Nickel Refinery that the company appears to have won). Implementing the scheme while Inco is in a strong position vis-à-vis the union probably means that it will not become an important point of negotiation and will meet little *official* resistance, but, as in CCNR, it may well become a direct point of struggle.

Developing alongside modular training is another major program in which jobs will be reclassified throughout Inco's operations. It is known as the Co-operative Wage System (CWS), has been implemented in Thompson, and is due for implementation in the Ontario division. The system originally evolved in the U.S. steel mills during the Second World War as a means to grant workers wage increases beyond the allowable limits. The steel companies and the Steelworkers collaborated to set up the wage evaluation scheme. Various factors, each with a different weight, are taken into account in arriving at the series of pay levels. Responsibility is given more weight than skill and skill more than hazards; physical effort is rated the lowest. In this way the hundreds of job categories are collapsed into a few pay levels, a goal similar to MTS's cutting of job classifications. Each job under CWS has a formally specified title, code, and classification in which the primary functions, tools and equipment, materials used, source of supervision, direction exercised, and detailed work procedures are set down as an agreement between the company and the unions.* According to a union official in favour of its implementation:

> *CWS is part of a bigger scheme. You have to be able to have some systematic way to use many incentives to get people to perform certain functions and behave in certain ways and develop certain work patterns and behaviour patterns. CWS is one of those ways of doing it. By manipulating wages you encourage people to perform or discourage people from performing in certain sorts of ways that you predetermine are the best for your method of production.*

The implementation of the CWS and MTS programs will have important implications for the mining industry. The company can utilize a worker's time more fully by training him in a

* Union officials are usually in favour of CWS because its implementation involves wage increments, as negotiated in collective bargaining agreements.

number of tasks, and he is then identifiable by the code numbers for these tasks. If he has any free time (determined by the amount of time allotted to any task), the company can require him to perform more tasks. What was once one job, done by one person, can be readily united with a number of other jobs, so that one worker finds himself doing all these tasks. To date the two schemes have not been in effect together at Inco, but it is evident that they will give the company a more accurate inventory not only of their equipment but also of the people capable of operating and maintaining that equipment. CWS will provide a systematic way to evaluate the wage levels for each of the modular tasks and the means whereby these tasks can be amalgamated.

Inco is obviously moving ahead rapidly in the area of "people technology" just as it has in other forms of technology. The efficacy of the training system remains to be seen, particularly underground, and the full implications of the two schemes are not yet known, but together they should give Inco greater control and standardize the work, adding further to the growing rationalization. The extent of the workers' resistance to these programs will be important to future developments in the organization of work in the mining industry. To date the unions have co-operated closely with CWS and have had little to do with MTS, except where it violates the CBA provisions concerning operators' being asked to do maintenance work. This is surely a source of future tension.

Contrary to the popular notion that increased technology leads to greater skill requirements, automation and mechanization in mining have on the whole had the opposite effect. In part this is attributable to the fact that machinery to some extent controls itself, but it is also the result of simultaneous changes in the organization of work and in the way workers are trained. James Bright of the Harvard School of Business identified this trend in 1958, arguing that "we tend to confuse the maintenance and design problems or exceptional operator jobs with the most common situation: namely, that *growing automaticity tends to simplify operator duties*."[19] Increased skill is indeed required for designing equipment and in some cases for its maintenance, but in the Canadian mining industry much of the equipment is designed outside the country (see page 347). Thus the potential benefits for the skilled component of the Canadian labour force are reduced.

While the stature of traditional craftsmen underground, responsible for entire work cycles, was being reduced by mechanization and that of skilled operators on surface in milling and refining was being reduced by automation, another category of worker emerged to maintain the equipment. In it are skilled tradesmen and engineers. They do not have the skills of the older craftsmen, but they are nonetheless skilled. Elaborate educational systems and apprenticeships have been developed to transmit these skills. This status has given these workers considerable power, reflected in their higher wages and, in the case of maintenance workers, their power within the unions. Much of their leverage has come from their freedom to change employers when there were shortages of skilled workers. But recent developments are threatening their skills. With the development of MTS programs for maintenance workers, individuals become tied to specific equipment and specific companies. Tradesmen become more expendable as companies develop means to teach limited aspects of their trades rapidly to unskilled workers. Nor does increased technology lead to a simple, expanding need for engineers. As we saw earlier, the amount of engineering time required for blast-hole mining is only one-fifth that needed for conventional methods.

We can see the increasing importance of maintenance over production in the mines and surface plants at Thompson, where maintenance is a separate department, in Appendix XIII. Between 1966 and 1977 maintenance as a proportion of the hourly labour force increased from 16 to 24 per cent (before 1966 maintenance had been higher than 16 per cent because of start-up requirements). Expressed differently, the ratio of production workers to maintenance workers dropped from 5.1:1 to 3.3:1. Although their proportion of the labour force is clear, there is some doubt about whether they will remain a scarce source of labour. The forces of change in training suggest strongly that there are important internal adjustments in this field also. MTS types of training will certainly continue to cover maintenance tasks, but these tasks are increasingly being performed either by operators or by workers trained to maintain specific equipment rather than by tradesmen. Since fewer maintenance workers will be tradesmen, there is likely to be a fall in the value of the labour power of maintenance workers and a diminishing of apprenticeships.

Another aspect of technological change with implications for workers' skills is the introduction of process technology and automation in surface installations such as the Clarabelle Mill and the Copper Cliff Nickel Refinery. Developments here are similar to those noted by Robert Blauner for chemical process technology: "Whereas continuous-process production increases the worker's breadth of knowledge and awareness of the factory as a totality, it does not necessarily enhance the worker's depth of understanding of the technical and scientific processes."[20] An operator who is associated with a broader portion of the production process is not bound to have greater knowledge of the equipment. Indeed, operators become the policemen of the system, reporting problems as they patrol the plant instead of actually performing the operation of the machines. The extent to which they operate machinery is in fact dependent upon directions received from centralized control rooms, so that their influence is further reduced and appropriated to centralized direction. Obviously the operator in these plants has a cleaner (although not necessarily safer) working environment and uses less physical effort. He possibly has greater "responsibility" because his activities cover a broader range and any action may have multiplied effects, but it is not the kind of responsibility of, say, a skimmer or furnace operator in the smelters. Nor does he have the same kind of control over the work process.

Does all this mean that operators of machines are less skilled? This question requires a definition of skill. Bright suggests that skill is "an indefinite blending of several things – manual dexterity, knowledge of the art, knowledge of the theory, and comprehension and decision-making ability based upon experience."[21] If we use this definition, operators in automated surface operations or mechanized mines have less skill than those in more traditional settings. The direction of change parallels that found by Katherine Stone in the steel industry: "Just as the authority that the skilled workers had previously possessed was transferred to the foremen, their overall knowledge about production was transferred to the managers."[22]

What has been the reaction of union leaders to these developments? Generally they have accepted them while at the same time trying to maximize the gains for their members, although not clearly articulating how these developments may provide gains for workers. The following comments come from two Sudbury union leaders:

> *I don't think we're totally opposed to technological change. We're opposed to the philosophy that you introduce technological change to replace workers. We want our people to run all this machinery, receive higher rates of pay. We can't stop the company from saying, "We're going to have 11,000 workers," because we don't have the control to say to the company, "Don't do that." The residual rights are still there.*

> *As for the introduction of new technology, I don't think we want to have a say in that. That is their* [the company's] *side. I think that they understand the importance of efficiency, and they are not going to introduce something that is backward. They are going to introduce something that is efficient, and a lot of times it's a labour-saving device. Anyway, it's a fight we wouldn't have a hope in hell of winning.*

Here is a passive approach to technological change. There is some resistance, to be sure, but it tends to be confined to operators' objections to maintenance work and the short-term displacement of workers by technological change. There is some resistance to the erosion of skills, but again this tends to be rather narrowly formulated and easily absorbed by management.

For the most part management has been successful in implementing changes in the techniques of production and training. In both areas they serve the twin goals of increasing management's ability to accumulate capital and social control of workers. The strategies have been costly; tremendous amounts of money have been invested in capital equipment and training programs, not to mention some wage concessions to workers. But in the long term, management feels these investments will increase their power. There is every reason to believe they are right.

Mechanization underground and automation in surface operations have had the most dramatic and far-reaching effects on the nature of the mining labour force. The effects have not been felt evenly in all operations because of the great range of methods we have seen. Nevertheless, the direction is clear: the new mines are mechanized and the new surface operations are automated. Parallel with changes in the equipment side of technology are changes in the skill or training side. Modular training is spreading like wild fire through the Ontario division.* Since it first appeared at Inco in 1970 the company has decided to expand this method to virtually all jobs. Its intimate tie with new equipment is readily apparent. Before modules could be designed it was first necessary to do an inventory of *all* the equipment and processes in mines and surface operations. These, in turn, had to be fitted into the entire stream or flow of the company. Each task was then specified and training manuals written on how to do each of the tasks or operate each piece of equipment. Unlike traditional apprenticeship systems or "trade" training (which the union is promoting), the company's module system is designed to minimize the amount of time needed to teach a new worker a new task. The worker does not learn the principles and tools of the trade but is tied to highly specialized equipment. He is thus less valuable on the job market because of this low skill specialization. Moreover, the worker is tied to the equipment, and his limits are those of the machine. Standards are measured and set and his performance monitored. The element of judgement and the value of experience are minimized. A raise borer can learn in a few weeks how to operate a machine and perform the same task as

* In February, 1980, Inco's Manitoba division announced implementation of a Functional Training Program for its surface operations, including thirty-eight modules for the mill, fifty-five for the smelter, and fifty-six for the refinery (*IN Manitoba* 10, no. 1 (1980):13).

a raise crew – traditionally the most skilled and respected of the miners, who needed years of experience to perfect their skills. The "craft" quality of production disappears, and the worker becomes a readily replaceable item. His replacement requires little more skill than a labourer; he becomes a machine tender or a machine monitor.

CHAPTER NINE

Fighting Back: Workers' Resistance

Boss man, boss man, what do you say?
Gonna get you alone in the mine some day.
I can't hold on but I can't let go,
And I can't say "yes" and I can't say "no."

Gordon Lightfoot
"Boss Man"

Individually, workers lose much of their control over the workplace with the advent of mechanization and automation. But the development of unions gives unionized workers, at least, greater collective strength to resist company strategies. This form of resistance has been highly developed in mining and other resource industries such as forest products where workers are concentrated in one-industry towns. Besides this solidarity they have had the benefits of work organized into "primary" work groups with considerable autonomy from supervision and consequently strong class loyalties. As already suggested, the miners' militancy has been heightened by management's actions with respect to layoffs and provoking strikes. For many miners, fighting back has become a way of life.

The freedom to form a union and the freedom to strike are

not natural rights; they are the products of long, difficult, and often bloody struggles by the working class. Strikes and unions are "rights" gained only through political pressure, usually against the resistance of capital. Even now they are not universal rights in Canada but are granted only under particular circumstances and to limited groups, excluding "essential services," and are often suspended by the state by means of periods of compulsory arbitration, back-to-work orders, and even, at times, force. Democracy stops at the plant gate. Unions in Canada face many roadblocks. They lack legitimacy, and this impediment is reinforced by the media's attitude toward them. It is instructive to note the emphasis the press places on time lost through strikes when actually more man-days are lost as a result of injuries – a fact seldom brought to the public's attention.

Canadian unions lack more than total acceptance. As a body they are divided in several aspects. About 60 per cent of the unionized workers are affiliated with international unions, 28 per cent with national unions, and 12 per cent with government employee associations. Unions are divided into "blue-collar" and "white-collar" groups; the internationals tend to be blue collar, such as the Steelworkers and Automobile Workers, and the nationals tend to be white collar, as are public employee unions such as the Canadian Union of Public Employees and the Public Service Alliance. French-language and English-language unions may belong to different parent organizations. Moreover, Canadian union membership is concentrated in certain regions, with 82 per cent located in three provinces, Ontario, Quebec, and British Columbia. As would be expected, most women members belong to the white-collar unions and make up 42 per cent of the membership of national unions but only 15 per cent of that of internationals.* Among the international unions the interests of Canadian workers may differ from those of their U.S. brethren.

* In 1972 only 14 per cent of the executive positions in national and 4 per cent in international unions in Canada were held by women, a large under-representation of their membership in both cases. As Pat Marchak has shown, unionized women become alienated in male-dominated unions and turn away from their original union support after they have been in the union for some time ("Women Workers and White Collar Unions," *Canadian Review of Sociology and Anthropology*, Special Edition, 1974:187-200).

Unions are certainly not the only answer to the problems of the working class, but they are one expression of its resistance to capital and an attempt to protect itself collectively. They have, however, been more successful in satisfying "economistic" demands for greater income than demands for control over the workplace, an area where the battle has been mainly a rearguard action. Moreover, unions represent only a third of the labour force, and this, in turn, acts as a potentially divisive factor within the working class. Some workers are alienated from their own unions. Unions often become, at the shop-floor level, agents of social control and discipline, being forced to carry out the collective bargains struck with management. Thus sit-ins and wildcat walkouts are often directed at the unions' inability to cope with workers' problems. To some extent the unions become part of management's strategy for guaranteeing its prerogatives. At the very least unions shuttle uneasily between workers' resistance and managerial strategies. They are at one and the same time the most systematic and organized expression of this resistance and, through the commitments they make to companies when they enter into collective agreements, a containment of many forms of workers' resistance.

The Battle of the Unions at Inco

Before the turn of the century, labour shortages and lack of organization were the normal situation in Sudbury. In his review of the period, Gilbert Stelter says: "The lack of organization made strikes an ineffective weapon. In what appears to be the only strike in Sudbury in the 19th century, employees at four of Canadian Copper's mines struck for higher wages in July, 1899 . . . [but] the strike 'fizzled out' almost as quickly as it began. There was no organization among the men, and the only result has been that a few of the ring-leaders have been discharged and most of the others have gone back to work."[1] Abortive attempts to organize Sudbury miners began in 1933 under the Mine Workers' Union of Canada but collapsed in 1934. This union later affiliated with the International Union of Mine, Mill and Smelter Workers (Mine-Mill) which began organizing at Inco in 1936. John Lang writes in his study of this union: "The organizational drive was given a shot in the arm with the success of the United Steelworkers of America signing a contract with INCO at Huntington, West Virginia on April 21, 1937. INCO countered this event with a 10% increase in wages as well as with plans to form a company union."[2]

A shortage of union funds, however, soon crippled the Mine-Mill campaign, and many preliminary gains were lost. In 1942 Inco attempted to forestall unionization by offering a contract to its own Employees' Welfare Association, which was in existence from 1940 to 1944 in Port Colborne and Sudbury. "A handful of members of this association in Sudbury, at a meeting held in the office of INCO vice-president Donald McAskil, constituted themselves a union – the United Copper-Nickel Workers (U.C.N.W.). Local 598 [Mine-Mill] reported that INCO officials working on company time were attempting to sign up members by offering them better jobs. The U.C.N.W., or 'Nickel Rash' as it was less euphemistically referred to by Mine-Mill members, masqueraded as a *bona fide* trade union by establishing an office off INCO property, by releasing press statements and by providing 'stewards' in a few plants. However, the U.C.N.W. was financed by INCO at a cost of $62,000 in 1942."[3] Mine-Mill faced formidable opposition from the company union, the church, the media, and the company itself, including goon squads hired to break up the union office and frighten its officers.[4]

An old miner recalled for us the days when Inco was being organized and some of the differences the union made in the workplace:

> *Mine-Mill tried to organize a union around 1936–37. They had Local 239 chartered for a while, but as fast as any of the bosses found somebody talking or discussing unions, they were simply fired, instantly. A chum of mine went out from the refuge station after lunch one day – we always ate in the refuge station – he went out from the refuge station this day, and at one of the corners there was a water fountain, just a pipe set up from the ground, and water was running there so we could get a drink. And there were half a dozen men standing around this water*

fountain as he came up and he could hear what they were saying, and they were saying something about unions. Well, he had four small kids to look after, and he knew the score, so he just turned and went on working. He never saw one of those men again. By the time he came back to the refuge station at quitting time those men had all been fired, sent to surface. They'd collected their pay, cleaned out their lockers, and were gone.

Would that be for talking union only at the workplace or anywhere?

Anywhere, for that matter, but particularly on the job. It was rumoured that they sent a few spies around to houses if they got some word that somebody had something to do with the union. All you had to do was join. They would send somebody around to the house and look around and see if he had any union literature lying around. If he did, he was fired when he went to work the next day. You see, it was the arrogance that really ate up the guys, and the one thing the ordinary man couldn't explain, or put into words, was the idea that the boss was making all the decisions; then when something went wrong the company looked around to see which foreman they could pin the blame on. So the man got fired and the foreman got fired too, or else demoted, at least, down to an hourly paid job again. During the war they had to change that. Well, by the time the war was over the Ontario Labour Relations Act had been passed and the federal Labour Relations Act, and the unions had a grievance procedure which is a compulsory part of every collective bargaining agreement and was written out right in the Ontario Labour Relations Act.

After the war, with a union there, every time a man had a grievance or was chewed at or anything like that, or had an unjust penalty, there was a grievance put in; and then there was a long grievance procedure over it which quite often went to arbitration. It didn't do much good to go to arbitration at that time because the judges who we had for chairmen of the arbitration boards were very pro-company and anti-labour. . . . But gradually these darned grievances were such a pain in the neck that the foreman tried to get along without grievance, and it gradually reached the point where, if possible, the foreman would try to settle the complaint without a grievance and the layoff penalties unless some higher-up caught a man. The boss would say, "Now look here, no more of that. You know the rules, and just do it the way it is supposed to be done," and then he would forget about it. But if some of the higher-ups or the safety engineers found a man doing something he wasn't supposed to, well then it would have to be a penalty, but it got so the penalties were reduced to just warnings at first and then a layoff penalty for any further infractions of the rules.

In August 1943, Mine-Mill for the third time in five years embarked on an attempt at certification before the Ontario Labour Court to oust the UCNW. In a vote at Inco, 88 per cent of the workers took part, voting 6,913 for Local 598 of Mine-Mill, 1,187 for UCNW, and 675 for no union. At Falconbridge workers voted by a similar margin to support Mine-Mill and Local 598 was certified. Soon Inco workers at Port Colborne were chartered as Mine-Mill Local 637. For the first time jobs were posted and a seniority principle implemented, challenging the arbitrary power of supervisors.

During the mid 1940s Mine-Mill was active throughout northern Ontario and Quebec. The leadership of the Canadian Congress of Labour (CCL), however, took exception "to the political ideology of some Mine-Mill officers."[5] In December 1945, after the war's end, while the government was still regulating wages, Local 598 could make only limited gains, but by June 1 of the next year it had "won an increase of ten cents an hour on a forty-eight hour week, the Rand formula for the check-off of union dues, important job reclassifications, and improved vacation provisions." John Lang concludes that "the problems facing Mine-Mill were not those of organizing and negotiating, but rather concerned the political battles within the International Executive board."[6]

In 1948 the CCL leadership began a campaign to oust Mine-Mill, moving formally for suspension in August and for expulsion in October. The next month the United Steelworkers of

America applied for Mine-Mill's jurisdiction. Thus Mine-Mill was confronted not only by the companies but by the forces of organized labour. The CCL, says Irving Abella, was "caught up in the frenzy of the Cold War . . . [and] sought to cleanse itself by ousting its left-wing unions. Aided by the greed of some unions anxious to take over Mine-Mill's jurisdiction and by the desire of some partisans of the CCF to exorcise their left-wing opposition, the Congress had little difficulty in expelling Mine-Mill."[7]

Vulnerable to raids by the Steelworkers, Mine-Mill also had internal problems. There was a confrontation over "national autonomy" in which three locals withheld dues; this caused another crack in Mine-Mill solidarity and a further opening for the Steelworkers. The confusion provided Inco with the chance to fragment the union further by withholding dues and refusing to bargain with Local 637 (Port Colborne). Local 598 (Sudbury) lent support and refused to negotiate with Inco unless Local 637 was recognized. At the negotiating meeting in April, 1949, Inco refused to recognize Local 637 and the international president, John Clark. In Lang's opinion, "the interests of the raiders and the company were complementary."[8] Following a legal battle in the Ontario Supreme Court, Mine-Mill won a judgement and was awarded the dues Inco had withheld "except for INCO's legal fees which were to be deducted from the fund."[9] Raids on Local 598 continued, hampering the union's ability to negotiate. This time the raiders were the United Mine Workers of America. In Port Colborne the Rand check-off formula was dropped by Inco; this "was of direct benefit to the C.C.L. and their supporters. Once again, the raiders and the companies were working hand in hand," according to Lang.[10]

In 1950 the Steelworkers again began to raid Sudbury and Port Colborne but without success. In the Timmins area, however, the Steelworkers won all eleven local mines. When there was a lull in the raiding, internal leadership struggles again appeared at Mine-Mill.

The Canadian Mine-Mill Council, a Canadian conference, was organized in 1953 in response to the autonomy issue, and in 1955 the international union granted autonomy to its Canadian membership.[11] Yet home rule led to further trouble. In a hearing before the Ontario Labour Relations Board, where Mine-Mill was requesting certification for a union at the Algom Uranium Mines in Elliot Lake, the board dismissed the case on the basis of "the Steelworkers' argument against Mine-Mill's certification. Steel pressed the point that Canadian autonomy meant Mine-Mill was a new legal entity and was therefore ineligible for certification."[12] Three more applications were dismissed for the same reasons. "Any existing contract with a local union was still legally binding, but the O.L.R.B. would not recognize Mine-Mill as a trade union until it re-signed a majority of the workers in each of the plants, and re-applied to the O.L.R.B. for certification. While this was being accomplished, each local union would be open to raids by the Steelworkers."[13] As a result of the rulings, Mine-Mill lost out to the Steelworkers in the uranium mines at Elliot Lake, despite the fact that the Mine-Mill local had signed a majority of the workers. These events foretold those soon to occur at Sudbury and Port Colborne.

In 1957 the metals industry was depressed, plagued by weak markets, overproduction, and low prices. By July 1958, twenty-one mines were closed and there were major cutbacks at Inco and Cominco, the two largest mining companies.[14] In March 1958, Inco laid off a thousand employees in Sudbury and three hundred in Port Colborne. At the second meeting of contract negotiators on April 15, Inco announced another layoff of three hundred employees. Negotiations were brought to a standstill, and a conciliation board was being established when a further 20 per cent cut in production was announced by Inco. "Rather than lay off the 2500 men that this decision required, the company decided to put all its hourly-rated employees on a 32-hour week. In effect, this amounted to an average wage cut of 55¢ an hour, reducing the net income of workers to their 1951 level."[15] The union was asking for a three-year contract and a 3 per cent increase each year above the average wage of $2.20 an hour. After the conciliation board recommended that there be no wage increase, 10,662 of the 12,887 Mine-Mill workers who voted called for strike action. When Inco was unwilling to negotiate, Local 598 struck on September 24, 1958, at a most opportune time for Inco. Not until December did Inco make an offer the union could accept, and this was 1 per cent the first year, 2 per cent the second year, and 3 per cent the third year. Production resumed on December 22. Meanwhile Inco had cut its inventories and saved three months' wages.

For the union leaders, the 1958 strike was disastrous. Tremendous community pressure was brought to bear against the strikers. On December 12, two thousand women attended a rally at a local arena led by the mayors of Sudbury and surrounding communities, reinforced by company officials. The purpose was to pressure the women to appeal to the strikers to return to work. Among the women were some strikers' wives told to attend the rally by their priests, and the press reported the fact as if the wives had turned against their husbands. The effect was demoralizing, and only ten days later the strikers agreed to the weak contract. But the outcome was the result of more than simply community and family pressures on the miners. From the outset the leadership had been indecisive. Initially they had recommended a strike; then they changed their minds, holding a meeting to recommend acceptance of the company's offer. By this time the membership was ready to go on strike. A former Mine-Mill official described the problems:

> *There had been a period of time before 1958 in which the leadership hadn't been concentrating all its energies on the job as they should have . . . and in 1958 there was an alienation between the leadership and the membership. The membership had very little faith in the leadership. During the strike I am quite sure the company realized there was an opportunity to give the union a real good shaking. They mobilized an awful lot of the community notables, so to speak. They got the wives to speak against the union. They were in a position of strength. When the men went back to work they went back for a few pennies more. The morale of the guys was completely broken. . . . At Falconbridge there wasn't a strike. Falconbridge workers traditionally stayed at work whereas Inco workers went out. In 1959 we had the next executive elections after the strike and a completely new board came in.*

Not only Mine-Mill leaders suffered in the 1958 strike; even Steelworkers leaders, who were the only beneficiaries in the labour movement, felt a great loss. Everyone was embittered. As one Steelworker who belonged to Mine-Mill at the time said, "I think the company, in the long run, lost the strike because people had a great determination, a great hatred for the company. That was a premeditated smashing of the union."

Meanwhile, turmoil continued in the internal politics of Mine-Mill. There was serious disagreement within Local 598 and between the local and national executives. Open warfare erupted at the Mine-Mill convention in September 1959. In January, 1960, the Canadian Labour Congress (CLC) turned down Mine-Mill's application for affiliation on the grounds it was "ineligible. . .due to the constitutional restrictions on communist organizations in the C.L.C. and the fact that the United Steelworkers of America held Mine-Mill's jurisdiction."[16]

Two strong factions had emerged by 1961 in Local 598, reflecting the tense relationship between the local and national offices. The so-called Gillis faction was making overtures to the Steelworkers (a condition the CLC had imposed for affiliation). The national Mine-Mill office placed Local 598 under trusteeship and occupied the union hall. When a struggle ensued, the Riot Act was read on August 27, 1961, at 5:30 A.M.

In September a rally at the Sudbury arena was attended by between six thousand and eight thousand men. There was an uproar, and the police broke up the meeting. By mid September "there was no question that Local 598's Executive Board was assisting the Steelworkers in the raid."[17] In mid November Inco refused to negotiate with Local 598. By this time the Steelworkers were mounting a concerted effort to take over Mine-Mill. In *Steel Labor*, the Steelworkers' newspaper, Larry Sefton, director of District 6, was arguing that Mine-Mill was being expelled first because of its communist affiliations and second because "it has not succeeded over the years in organizing the men in the nonferrous mining and smelting field."[18] Although the first claim was true, it had little justification. The Communist Party was legal in Canada. The second claim needs to be modified. Mine-Mill had been very successful in organizing not only in mining but also in other industries.

During the previous ten years, however, it had been constantly under attack by other unions for its politics, and its locals were being taken away by the actions of both the government and other unions. The real defeat of Mine-Mill at Inco can be attributed primarily to the strike in 1958 instigated by Inco, which brought the union to its knees. The raids had in fact been close-fought affairs.

In a vote at Port Colborne early in December, 1961, the workers chose the Steelworkers over Mine-Mill by a margin of 1,033 to 763. In Sudbury there were 14,333 eligible voters of whom 7,182 voted for Steel and 6,951 for Mine-Mill. This was a majority of 231 but a margin of only fifteen votes over the mandatory support of half the eligible voters. Because of allegations of forgery the Steelworkers withdrew their application at Falconbridge, but they retained the much larger Inco local. The OLRB granted certification on October 15, 1962.

In Thompson, Mine-Mill began organizing in 1958 and formed Local 1026, which was certified in June 1960. Before being organized, these miners were working at a base rate of $1.50 an hour on a fifty-four-hour week.[19] Mine-Mill was seeking the same contract that it had for Local 598 in Sudbury and twenty cents an hour isolation pay. A conciliation board proposed a ten-cent increase with another 10 per cent in the second and 11 per cent in the third year of a contract. The members of Local 1026 rejected this proposal, later accepting a company offer of another two cents an hour in the first year plus an irrevocable check-off. Inco apparently encouraged the formation of a Mine-Mill local in Thompson in an attempt to keep the Steelworkers out, but after Mine-Mill was ousted in Ontario, 84 per cent of Thompson's workers turned out in April 1962 to vote 1,226 to 352 to be represented by the Steelworkers.

Inco did not remain passive while all this union in-fighting was taking place. Lang's account says:

> When it appeared that a decision regarding the union dispute was at hand, INCO set the stage for bargaining by announcing a 13% cut-back in production (September 24, 1962), resulting in the lay off of 2,200 employees. When

> Steel was granted the bargaining rights at INCO, the company took the position that any conditions negotiated by Mine Mill would not be extended to the Steelworkers and proceeded to bargain from scratch. It was to be another eight months before INCO workers would enjoy the protection of a collective agreement. The most direct benefit that the Steel raid provided to INCO was that the company was able to operate for a nineteen-month period without providing any wage increases.[20]

In 1965 Mine-Mill attempted a counter raid on the Steelworkers at Inco but failed by a margin of two thousand votes. In April 1967, the national and international offices of Mine-Mill merged with the Steelworkers. In Sudbury, however, Local 598 representing the workers at Falconbridge rejected the merger. It continues to exist as an independent union.

The Steelworkers' struggle with Mine-Mill is all but over. Feelings in Sudbury continue to run strong among older miners, but there is evidence that the rift is no longer so large. The struggle did mark an important break in the philosophy of mine unions. Militant political unionism changed to a form of corporate unionism characteristic of the CIO–CLC. Labour unions in mining, as institutions, are now essentially business unions, even though there are many radical and progressive people in them. They are bound by contractual arrangements with private corporations and the state. Their legitimacy is fragile and granted by the state, to be withdrawn whenever they violate their "contract." Because of their position they have taken over part of the role of worker control. They represent a collective body for disciplining the work force and too often become formal bureaucratic channels for workers' grievances.

Since the Mine-Mill merger, the Steelworkers have come to dominate the labour movement in Canadian mining. Contracts between Local 6500, Canada's largest local until the 1978 layoffs, and Inco have set the pace for the entire industry. Table 23 shows the distribution of unions in Canada's mineral industry. In terms of membership in all sectors, the Steelworkers, with 200,000 members in Canada, ranks behind

Table 23

Trade Unions in Canada's Mineral Industry, 1974

	Membership	Per cent
United Steelworkers of America	60,898	75.3
Confederation of National Trade Unions (CNTU)*	6,000	7.4
United Mine Workers	5,800	7.2
Mine, Mill and Smelter Workers	3,438	4.3
Canadian Association of Mechanical, Industrial and Allied Workers*	2,500	3.1
Democratic Federation of Metal, Mine and Chemical Workers (CSD, Quebec only)	2,000	2.5
Canadian Mine Workers	220	0.3
Total	80,856	100

SOURCE: J. A. MacMillan, G. S. Gislason, and S. Lyon, *Human Resources in Canadian Mining* (Kingston: Queen's University Centre for Resource Studies, 1977), p. 94.
*Estimates.

only the Canadian Union of Public Employees, with 245,000, and well ahead of the United Automobile Workers, the next largest industrial union, with 135,000 members. In recent years the Steelworkers have been challenged by the Canadian Association of Mechanical, Industrial and Allied Workers, an independent national union based in western Canada, but it has gained only a few locals and about 2,500 members.

Most hourly paid workers in the mining industry are covered by collective agreements; as of October, 1976, 83 per cent of non-office employees were covered. Only 11 per cent of office employees in mining had a collective agreement at that time. This indicates why women constitute only about 2 per cent of the members of unions in the metal mines industry.[21] One of the few locals covering office and technical workers is Local 6855 of the Steelworkers at, ironically, Falconbridge. Formed in September 1966, it is the largest office and technical local in Canada with six hundred members, 15 per cent of them women. Included in the local are, in addition to the usual office and technical workers, "Technicians, Technologists, Engineers, Accountants, Assistant Paymaster, Planners, Programmers, Safety Inspectors, Incentive Administrators, Work Study Analysts, Chemists, Geologists, Surveyors" – 109 different occupations in all.[22] Two attempts by the Steelworkers to organize a similar local at Inco have been unsuccessful.

The collective agreement for Local 6166 in Thompson specifically excludes "office staff, engineering staff, geological staff, foremen, shift bosses."[23] Recently, however, the Canada Labour Act was changed, so that it is now possible in Manitoba for first-line supervisors to become part of a union. In Ontario, supervisors are still prohibited from union membership.

Local 6500, the cornerstone of the Steelworkers' mining membership, suffered in the seventies because of the cutbacks. Its membership declined from 18,000 in 1971 to 14,000 in 1977 and in March, 1979, stood at 11,100. Trends suggest its membership may be as low as 7,000 by the early 1980s. Declining membership is only one of the problems facing all three Steelworkers locals at Inco. During the 1978–79 strike, tensions became apparent among the three locals and between the locals and the national office.

Workers have generally made substantial gains through unions compared with their position under the pre-union arbitrary power of the companies. One such gain is the principle of bidding, which is based on seniority and a minimum training level. It allows individual workers to shift their work location and move up or down the job hierarchy. A veteran miner may, for example, wish to move from an incentive job to one with less physical strain or risk, if he can afford to, while a younger miner may wish to increase his income and bid to a higher-rated job. Bidding gives mobility to the individual without changing the structure of the work and in principle applies just criteria to "promotions," thus decreasing arbitrary power or favouritism. A worker may even be willing to remain for a time in a job he does not like if he sees a move in the future. On the other hand, bidding can be disruptive to the collective interests of workers when jobs are made redundant or there are layoffs and the senior worker may have to knock another worker out of a job or take a lower-rated job against his will. With bidding and the seniority principle unions become part of the structure to maintain order within the workplace. One union official told us: "The union itself has put more constraints on workers because it has built all these protections for them and it binds them to the job even more. A pension after thirty years rather than at sixty-five is that much closer if you are twenty. Ten years' seniority gets you some pretty good jobs. So you are bound that much closer to the job. We haven't done ourselves all that many favours, either, with job protection. I mean, you have to have it – there's no doubt about it. But it works against us in some areas."

One of the ironies of the labour movement is that companies are able to turn unions back on the workers and use the unions as instruments of social control. Another union official reinforced this observation: "Business unionism is a deliberate attempt to reduce the militancy within the union. They [the companies] don't want to see the union destroyed. They want to see the union do a specific job in which they can have complete control over it. They have an interest in having the union there." Once unions are in place, companies do not so much attempt to undermine their existence as to reduce their power. Unions are confined in a legal relationship that has little room for moral questions or injustices not covered by the CBA. Unions become buffers between the workers and the company and limit the forms of action possible to workers.

These inconsistencies are manifest within the union structure itself. Unionists differ among themselves about their relationship to the membership. The following exchange between two union executive members illustrates the debate over direct and delegated democracy:

> Officer 1. *We still have this idea that we elect eleven officers and give them the job of running the union; and every month they go to a meeting where maybe thirty guys show up, and those guys can dictate to the eleven elected officers. . . . There's nothing wrong with having a membership meeting for people that want to come and hear what the hell is going on and speaking their mind, but I don't think that they should be able to dictate to the people that are elected by the 14,000.*

> Officer 2. *But I disagree. I think that the members – even though not that many come – they're the ones that should run it. I don't think eleven officers should run it.*

Officer 1. *I think in a big local union you have 14,000 people with the opportunity to vote, and that's why they voted. They're selecting people to do the business.*

Officer 2. *Okay, I think that they should come to you and say, "Well, here's what we are going to do," and then you have the right to say, "Well, okay, I disagree or agree with you." But I don't think the officers should say, "Here, we are going to do this and we're going to do that if you like it or not."*

Officer 1. *But I don't think just because I can bring fifteen or twenty guys out to a meeting that I should be able to screw up the officers and dictate to the 14,000 members as well. I don't think that is right.*

More than management, and even more than government officials, union leaders are directly subject to the wishes of their membership. Major issues are frequently put to a vote at the regular monthly membership meetings (one held in the evening and repeated the next day for those on shift). Union leaders, as the evidence of strikes shows, are frequently placed in a no-man's land between outside authorities and their membership.

Strikes at Inco

Under a clause in the collective bargaining agreement, the company receives a guarantee against cessation of work that allows them to plan their production and protects them from unforeseen work stoppages, enforceable by law. A typical clause is that in the CBA for Local 6500: "Neither the Union nor any employee shall take part in or call or encourage any strike, sit-down, slow-down, or any suspension of work against the Company which shall in any way affect the operations of the Company, nor shall the Company engage in any lockout at its Plants or Mines."[24] It is not possible, of course, to prevent wildcat strikes, but the CBA clause does make the union responsible when they occur. They continue at Inco, especially at Thompson, but the company then uses them against the union at contract negotiations. The union is forced to make concessions as trade-offs for dropping suits or reinstating employees suspended during wildcats.

Table 24

Strikes and Lockouts and Their Duration in Mining, 1948 to 1975

(In existence during the year)

	Strikes & lockouts	Workers involved	Duration of man-days
1948	12	16,695	308,989
1949	15	8,737	507,636
1950	15	7,258	47,800
1951	23	19,189	146,969
1952	26	9,539	91,825
1953	27	15,274	631,918
1954	23	9,227	196,169
1955	9	2,092	17,185
1956	24	17,974	58,630
1957	23	18,084	165,772
1958	30	30,909	808,840*
1959	14	5,611	25,740
1960	18	4,806	20,780
1961	11	5,944	31,740
1962	10	7,688	41,040
1963	17	6,210	53,980
1964	12	6,560	69,640
1965	25	8,402	58,460
1966	36	43,990	450,340*
1967	24	7,084	32,050
1968	21	4,882	100,800
1969	27	31,511	2,087,490*
1970	15	6,876	53,680
1971	19	7,680	193,490
1972	32	13,410	334,680
1973	33	11,560	220,570
1974	59	n.a.	509,380
1975	45	n.a.	1,173,040*

*Major Inco strike years.

In Table 24 figures for all strikes and lockouts in the Canadian mining industry between 1948 and 1975 are brought together. It is evident that there is considerable variation in the man-days lost and that some of the peaks occurred during the main strikes at Inco in 1958, 1966, 1969, and 1975. The 1958 strike that broke Mine-Mill lasted ninety days.

The early Steelworkers' contracts did not make major breakthroughs in wages or working conditions. The membership was still divided, and there was little support for strong bargaining in 1963. "But come 1966 there was some pretty strong support for the union," one union leader said. With demand for nickel high because of the Vietnam War, the Steelworkers were in a position of strength. During negotiations, however, a wildcat began in July at Levack Mine, spreading to the Iron Ore Recovery Plant, and the next day virtually all Inco workers in Sudbury (13,000) and Port Colborne (2,000) were out. This wildcat strike lasted three weeks. By the time the contract expired in September the union was facing a multimillion-dollar lawsuit and firings. The company placed these trump cards on the negotiating table, decreasing the strength of the union's demands.

* * * * *

A Sudbury union leader on the 1966 strike:
We had an illegal strike, you might say. The strikers shut down the company's operations and locked all the supervisors and everybody in and the company tried to supply them by helicopter. There were special police and four thousand guys were on the picket lines. Everybody just walked off the job and left everything running. Furnaces froze up. The whole god-damned thing was plugged tight. It lasted about three weeks. There were some sixty guys discharged, and when they finally got that thing going, it took months to get operations back because it was just plugged tight.

* * * * *

The settlement included getting their jobs back for the discharged men, some gains in safety and contract "language" (that is, matters subject to grievance), and an increase of sixty-six cents over three years. The wildcat, according to a Sudbury miner, "started out with a slowdown because we were told negotiations weren't going well, and the company reacted and started to penalize the guys like crazy; so the guys walked out." The bargaining committee finally told the workers to go back to work or they would lose in the negotiations.

In 1969 over two million man-days were lost because of strikes and lockouts in the mining industry. The Steelworkers struck Inco for 128 days; 1,600 men were out in Port Colborne and 15,000 in Sudbury. The conditions were right for the union. This time they were organized; one union member said, "We had picket captains. You could pick up your vouchers on a certain day. Everything was organized. The guys were happy. We were off four months, and we went back by only fifty some votes." The slogan was No Contract, No Work! The company was in no position to sustain a long strike because the war demands for nickel were continuing. The miners made major gains, including a 35 per cent wage and benefit increase over three years.

* * * * *

A Sudbury union leader on the 1969 strike:
Between 1966 and 1969 we did an educational on our guys: you can't do the 1966 stuff because we don't want to be at the bargaining table in an illegal situation with police and the whole thing, and we are going to fight the company with total discipline, especially for a long strike. The company was preparing another 1958; the stockpile was three months. We knew it, but we also knew we had to get into a battle. We just couldn't take a licking; we would have to win it. . . . The company wouldn't even talk. Their history of bargaining, believe it or not, in 1969 was that they talked money. We were on strike three months before they agreed to sit down to talk language. The company went to the members by ads in the paper saying we offered this and that. They tried to get the women organized as in 1958. In that particular strike our resources didn't run out. . . . The men wanted to kick the shit out of the company. It became an issue. We won almost every issue we were after. It became an issue to teach them a lesson that they had forgot in the past.

* * * * *

Following the 1969 strike Inco replaced its negotiating team and began its new industrial relations program.

Conditions were different in 1972. Just before negotiations began, Inco laid off six thousand employees, or about a third of the labour force. The nickel market was down, but this time the company and union came to an agreement without a strike. The 1975 negotiations were more complex. Inco had built up about four months' inventory and demand was neither high nor low. The Steelworkers were facing strike possibilities in the steel industry as well in a few weeks (the United Steelworkers also represent employees of Stelco and several other steel manufacturers, such as Algoma Steel, whose contracts expire at about the same time as those at Inco). Negotiations dragged on until the last minute, and under the "no contract, no work" doctrine, the men in Sudbury walked out. But the company and the union then announced a tentative eleventh-hour agreement. Two rank-and-file members recount these events:

Miner 1. *They told us nothing about what was happening in the bargaining, and we had no information at all.*

Miner 2. *It was a sell-out. That's what it was.*

Miner 1. *It was. It was a sell-out.*

Miner 2. *The guys running the union just sold out.*

Miner 1. *Twelve o'clock midnight deadline came and still nothing, so we went on strike. Everybody was hepped up. It was summer, and there's all kinds of young guys saying, "Hey, let's have a strike. Three months of holidays. Let's go." And I wasn't home ten minutes, I had a beer in my hand. We were yahooing, and we turned on the TV and there was our union president. "Yes, boys, we got you a contract. Go back to work." He wouldn't tell us what it was. We went back to work anyway.*

After some delay the membership was persuaded to ratify the agreement and return to work.

The next strike in Sudbury was the Day of Protest against federal anti-inflation legislation on October 14, 1976. As a reprimand, Inco selected forty-five men, mostly active union members, and gave each one Step Four penalties. The major struggle against the government's wage control policy, however, occurred in Thompson. The October 14 protest there led to Step Four penalties for twelve union members. These were all active unionists, mainly from the executive. Later an arbitration board decided that the Day of Protest strike by

Local 6166 was not illegal and did not award Inco the $200,000 it claimed to have lost as a result of the strike.[25]

The first official strike in Thompson took place in August, 1964, and lasted twenty-eight days after a company offer of 38.6 cents was turned down by 964 to 186. Besides wages, the major issues in the strike were working conditions, the cost of living, and construction of a year-round road from the south into Thompson. The government agreed to construct the road and take action on profiteering by local landlords and storekeepers. One of the interesting results of the strike was a mass exodus from Thompson. It is reported that 2,500 people left immediately, including about half the 1,900-man Inco labour force. Many of those who left would have voted to reject the company offer.[26]

The next strike in Thompson was not until 1973, when there was a one-day walk-out in the smelter and maintenance departments over gas emissions. In 1974 there was another wildcat over wages and inflation, with workers asking for more money because of increases in the cost of living. Since negotiation of the 1973 contract, costs had increased over the limit provided by the cost-of-living-allowance (COLA) in the contract. This strike was led by maintenance workers, and the operation was closed down for four days with a thousand men on the picket lines.

* * * * *

A Thompson union official describing the 1974 wildcat strike:

What they did was walk out, go into the woods with chain saws, and cut down logs and build log cabins on the roads into the mine sites. They went to great lengths: they put bunk beds into them (the one at T-3 had a veranda) and made rocking-chairs and built a big shit-house with a porch on it. They got portable generators and polyethylene and put big roofs on them. We were prepared for months. They couldn't get the foremen in; some foremen were in the plants, of course, and they had cots; but they didn't have much grub, so they had to hire helicopters, until some guy called the radio station and said: "The next helicopter that goes over ain't comin' back." So then the helicopters quit. . . . It wasn't us [the union]. *I'm not going to threaten that, but some guy sitting in the bush with a .303 can knock a helicopter out of the sky with no problem at all.*

* * * * *

As a result of the strike, the length of service package mentioned earlier was introduced. It included an increment of about eighteen cents an hour overall plus COLA for a total of twenty-five cents an hour. The company fired one man, but he was reinstated after the discharge was grieved and won. One company action placed the union executive in a predicament, however, as a union leader said:

They launched a lawsuit – they couldn't get us back to work first of all, so they filed an injunction. They wouldn't negotiate and that was it. Of course, wildcats are very ticklish situations. The union doesn't want to be found liable, but they had to deal with somebody, so they wanted to talk with the executive. We talked with them; but they didn't want to negotiate, so there wasn't much we could do. But then they got the injunction. They held the union responsible and served an injunction on the president and told him he had better get the men back to work some way or another. Subsequently they brought a lawsuit against us, but they couldn't pin it on us. You have to remember the union is risking two things. First, there is the lawsuit, but the union also risks being undermined as the official body. Now, if that injunction is served on one of the wildcatters on the picket line, he essentially becomes the leader of the union, doesn't he? And that union is no longer effective with a leader all of a sudden emerging from the rank and file.

This was not the last strike in which the union executive in Thompson was placed in a difficult position by the threat of injunction.

In February, 1976, Local 6166 voted to accept a contract of 18.5 per cent in the first year, 7.3 per cent in the second, and 6.3 per cent in the final year. In the Thompson case, however, the Anti-Inflation Board (AIB) rolled back the increase to 12.9, 8, and 6 per cent respectively. Under appeal the roll-back was reduced to only 14.9 per cent in the first year. The workers decided to challenge this decision, and 2,800 members went on strike on June 1. This conflict came at an inopportune time for the union, as an official recounted:

> *The union had just gone through a hard-fought election, and a new slate of officers had just come in but were in a state of limbo. There was a new executive going in and an old executive going out, a new president coming in and an old president going out. But the new executive did not come in legally until June 15 at that membership meeting. So there was no union leadership at the time. The town was stirred into a hysteria. It was just a powder keg. And the newspaper and the radio station had stirred up so much sentiment against the union. Basically, their question was, "What's the use of striking the government? We'll never recover. The businesses will go bankrupt; you'll lose your jobs, your homes." And they managed to split our membership right down the middle. But on June 1 they struck at midnight.*

The new president was hastily installed, along with the executive. Manitoba Premier Ed Schreyer asked the workers to delay the strike for fifteen days for a direct appeal to the Cabinet in Ottawa. Although the union executive council of sixty members agreed to the extension, the rank and file, according to an official, "probably quite rightly said, 'You have not put it to a vote, and we are not obeying it.' " A meeting was therefore called at the baseball park, and the membership agreed, by a narrow margin, to extend the deadline to June 15. Some strikers disagreed and remained on the picket line. By this time the company was threatening an injunction, and the Canadian Association of Mechanical, Industrial and Allied Workers appeared as a potential raiding union. The executive felt obliged to break up the picket line. One official said, "Eventually we tore the pickets down, a very unusual job for a president and his executive."

Representatives, including Premier Schreyer, Dick Martin

(president of Local 6166), Lynn Williams (director of District 6, USWA), Len Stevens (director of District 3, USWA), and several company representatives then went to Ottawa. They talked with Donald Macdonald from Cabinet and then with AIB official Donald Tansley, who went to Thompson. Both the Steelworkers and Inco sought to have the original agreement re-instituted. Inco was willing to pay the higher wages. Maintaining employment in Thompson was difficult at the best of times, but in the past year there had been a 118 per cent turnover in the town so that Inco had had to hire 3,600 new employees to keep a labour force of 3,000. June 15 came and there was no settlement. The workers struck again, this time for thirteen days. On June 28 the membership of Local 6166 accepted Tansley's recommendation of 15, 9.8 and 6 per cent over the contract. The union felt this was a victory because they eventually got everything back; what they lost in the first year was made up in the second. They had beaten the AIB. Moreover, this show of solidarity was spontaneous. Although the day-to-day business of the union does not involve many members and the monthly membership meetings are poorly attended in Thompson, it is apparent that, when called upon, these workers are ready, willing, and able to resist.

The next round of negotiations at Inco took place for contracts terminating in September, 1978, in the Ontario division and in March, 1979, in Thompson. The conditions for these negotiations on the company side paralleled those of 1958 and 1972: high inventories and large layoffs prior to negotiations. The situation for all the union locals was obviously a difficult one. They were in no position to make large demands, but the cost of living continued to escalate and cut into the workers' real wages. They had to fight just to stay even. Local 6500 led the way.

The longest strike in Inco's history began on September 15, 1978, when the membership of Local 6500 rejected by 6,319 to 4,141 (61 per cent to 39 per cent) a company offer. The base rate in June, before the offer was made, was six dollars an hour and seventy-nine cents COLA. Inco proposed to extend the old contract for one year with no increases. In August it offered the same base rate but with an increase in the COLA beginning in November. Before a vote was taken, Inco agreed to an across-the-board increase of ten cents an hour but deducted six cents from the earlier COLA offer. In spite of the odds, the membership strongly rejected the company's position as an insult.*

Yet the initial deadlock issue in the 1978–79 strike was not so much wages or benefits as the grievance procedure. One victory of the 128-day strike in 1969, when the union held the upper hand because of the high demand for nickel accompanying the Vietnam War, had been an effective grievance process by which union stewards were to be present at any final step proceedings, held at the Copper Cliff headquarters. These stewards were released from their normal work to attend and often met at the union hall before the hearings to plan strategy. Now that the company was in a strong position, Inco negotiators tried to use the 1978–79 talks to recover ground, insisting that the final step hearings should take place at the point of the initial grievance and that the number of outside stewards that could attend should be reduced. This would mean a weakening of the union's ability to represent and protect its members because the hearings would be held in the plant or mine manager's domain and there would be less opportunity for strategy. Decentralization of the hearings would water down the union's resources, already depleted by a drop in membership of nearly 40 per cent. The union offered a compromise giving twelve of its chief stewards one day off each month to plan strategy and the other three hundred stewards three days a year, but the company rejected the proposal.

Tensions among the union executive appeared. Stewart Cooke, director of District 6 of the United Steelworkers, spoke against the recommendations of the bargaining committee of Local 6500 in Sudbury and Local 6200 in Port Colborne (which negotiated together). The presidents of both locals, Dave Patterson and Ray Moreau, reacted strongly against what they considered Cooke's interference. Cooke had met Premier William Davis and J. Edwin Carter, chairman of Inco, on September 15 in a last-minute effort to prevent the strike, but their joint proposal was rejected by the bargaining committee. Speaking from Atlantic City, New Jersey, Cooke by-passed the

* A Sudbury miner is reported to have said, "We had no choice but to vote strike. The Company's offer was an insult. To accept would have humiliated us and broken the union" (*Globe and Mail*, 4 Sept. 1978:9).

negotiating committee and went directly to the membership: "I know of no way we could expect a change in the company's bargaining position for quite some time. Therefore, I recommend that members accept the offer and do not go on strike." His view apparently reflected the position of the union's international headquarters in Pittsburgh.[27] Only one member of the negotiating committee had voted to accept the proposal. The pressure was increased by the stand of former Ontario New Democratic Party leader Stephen Lewis against the strike vote in an article in the *Toronto Sunday Star* of September 17. Lewis contended that the Sudbury local "has fallen into militant and often unpredictable hands." This charge was rejected by the NDP executive.[28]

The Sudbury workers rejected the proposal, but in Port Colborne the membership decided to accept the company offer even though the president, vice-president, and negotiating committee had rejected the one-year contract. The president and vice-president then resigned their offices. Following the 1978 summer layoffs, the local had been reduced to a mere 750 members. The plant, redundant since the building of the Copper Cliff Nickel Refinery, had become essentially a warehouse where the stockpiled nickel was stored. The members had given up hope of being able to fight Inco and sought to hold onto their jobs as long as possible.

Throughout the strike Inco refrained from laying off the 750 hourly workers and 140 staff in Port Colborne and the 1,800 staff in Sudbury. In Sudbury, industries dependent on Inco laid off workers. CP Rail and CN Rail laid off fifty workers about a week after the strike began, and Canadian Industries Limited, which operates the sulphuric acid plants adjacent to Inco, laid off ninety workers. Inco's Shebandowan operation north of Sudbury laid off 215 hourly workers on November 3 but kept on all thirty-six staff.

At the outset of the strike pickets refused to let staff and supervisors into the plants and mines. The company countered with the use of helicopters and threats of legal action; picket trailers were not allowed to use hydro sources. Eventually terms of access were negotiated, allowing staff and supervisors through the lines. Dave Patterson is reported to have said, "We were afraid of a violent confrontation with the police and with company supervisors if they tried to charge the lines."[29] Everyone who passed through the picket line had to show identification to the picket captains and enter on foot. Generally the strikers were disciplined, but on February 9, 1979, after talks broke down, pickets blocked the entrance to Inco's plants and mines for four hours.

We saw earlier that Inco's position prior to the strike was one of heavy debt, amounting to about $1 billion, incurred in the Indonesian and Guatemalan operations. Moreover, profits were down from $73.6 million in the first six months of 1977 to $57 million in the same period in 1978, and the company wished to decrease its debt. One way to do this was to improve its cash flow. This could be done by selling the stockpile of nickel and not paying wages, thus freeing cash. During the strike Inco continued to pay nearly three thousand salaried workers, but it saved millions on the wages of hourly workers and supplies. There was a limit, however, to the advantages of the strike for the company. During the strike it could not produce copper and precious metals, for which Sudbury is the primary source, and the market for these metals was very high. The nickel market also began to turn around early in 1979, and the stockpiles were substantially reduced. After about six months it was in the company's interest to return to work. During this period the workers were striking the nickel stockpiles (a *de facto* lockout by the company); after that time they were striking Inco.

Management underestimated the Sudbury miners' resentment and will to fight. The miners were unwilling to negotiate away earlier gains and refused to allow the company to destroy the power of the local. Had they settled in September, 1978, they would have lost union strength and still have faced the stockpile with only a one-year contract while the company hinted further layoffs were likely. A rank-and-file Inco worker's letter to the editor expressed the miners' opinion of the company: "They had misread the workers' will to fight for a decent living wage. . . . The whole exercise was designed to lower the pride of the ordinary workers, to teach the pensioners a lesson for daring to picket Inco's Toronto office and to restrict the effectiveness of the stewards to obtain redress for aggrieved workers."[30]

The strike became a symbol of resistance to the erosion of union strength. The local president said, "I compare this [strike] to the right to organize."[31] Even the *Financial Post* acknowledged its importance: "For the labour movement, the strike could become an historic landmark – one of those events that not only are seen as turning points, but give labor its mythology, if not its soul."[32] Unlike the 1958 strike, this one received support from the community as the townspeople, including the shopkeepers and many local politicians, rallied behind the workers against the company. The national labour movement came out in strong support.

At the beginning of the strike Local 6500 had a million dollars in its strike fund and was entitled to thirty dollars a week from the international union for each member. The money from the international covered the strike pay of twenty-five dollars a week for each single worker and thirty dollars for married workers, plus an allowance of three dollars a child. But the local had to find $550,000 a month to pay the strikers' medical (OHIP) and life insurance premiums. A drug committee was struck to pay members' emergency medical bills and prescriptions as well as baby food and formula at a cost to the local of $9,000 a week. The cost to the local of running the strike was high because it paid for benefits normally covered by the company.

In a show of solidarity, the Mine-Mill executive were among the first to offer assistance. Despite the traditional rivalry, Jack Gignac, president of Mine-Mill, is reported to have said, "The response of our members in support of the Steelworkers is a natural response that workers have toward one another when they are in a fight. We have to do everything that is humanly possible to assist in building the unity that is necessary."[33] The membership were somewhat less enthusiastic, rejecting a proposal to double their union dues and turn the extra $21,000 over to the strikers, but they did approve a $10,000 donation and allowed Steelworkers to collect at Falconbridge plant gates twice a month. Throughout the country workers and unions made donations. Steelworkers' locals made initial donations at the outset (1005 at Stelco gave $60,000); United Automobile Workers' locals, locals of the Canadian Union of Public Employees, and many others contributed. In addition, many workers gave directly out of their pockets. The Canadian Union of Postal Workers, for example, gave $10,700 collected from individual members, not from union funds. Fund-raising benefits were staged in Ottawa, Toronto, and Hamilton, and frequent speakers' forums collected donations.

One of the most significant expressions of solidarity in the light of the 1958 strike was the formation of the Wives Supporting the Strike committee. An organizer said at the time, "We believe the sooner we get organized, the less chance there will be for that kind of manipulation to take place this time. We know that if we are supporting our husbands, the men will be able to stay out long enough to get a decent contract."[34] The main purpose of the committee was to fight the psychological battles of the strike. They organized a Christmas party and received donations of toys, including $35,000 worth from the St. Catharines and District Labour Council; they established crisis centres to deal with marital problems; they collected over $350,000 in cash to support the strike; they distributed "Stretch Your Nickel" cook books; and they held "bean dinners" for strikers and their families with the theme "The Company Gets the Gravy and the Workers Get the Beans." This committee of some two hundred volunteers became the backbone of the psychological war, with help from the Citizens' Strike Support Committee made up of local business people who assisted with donations and by extending lines of credit.

No break occurred in Sudbury for 165 days until the contract for Local 6166 in Thompson came due. At Thompson Inco offered a thirty-and-a-half-month contract beginning March 1, 1979, with a thirty-five-cent increase the first year and ten cents a year for the next two years (the same offer made to the Sudbury local during the strike) plus COLA to be paid quarterly, beginning July 1, 1979. The Sudbury workers had been offered a COLA provision coming into effect in May, 1980, with the first payment in September, 1980, and only two other assured payments (in December, 1980, and in March, 1981). The COLA provisions made the Thompson offer more attractive; in addition, no changes in the grievance procedures were suggested as at Sudbury, and pensions were not as serious an issue. There was also some confusion among the Thompson executive, who understood that the same offer would be made in Sudbury and accepted by Local 6500's executive. This turned out not to be the case, but before the matter was clarified the vote was already in, and 60 per cent of the membership were in favour of the agreement (an election shortly thereafter ousted the executive). Members of Local 6166 accepted the offer and did not go on strike – the membership of Local 6500 had done their striking for them.

After Thompson accepted, greater pressure was placed on the Sudbury executive to reach a settlement. But now the shoe was on the other foot, as far as the Sudbury membership were concerned. They felt they had a great deal of lost ground to recover and were willing to continue the strike to do so. They had weakened the stockpile; nickel markets were again strong; and the threat from off-shore nickel production proved to be less significant then they had been led to believe. The odds had been narrowed. They had gained strength and conviction, not to mention hostile resentment, through their confrontation.

Pressure came from several quarters besides Inco. Prime Minister Trudeau told the Sudbury strikers on March 9, 1979, "to negotiate with some good sense." This was in response to a question from John Rodriguez in the House of Commons revealing that Inco had contributed $20,500 to the Liberals in 1977. Closer to home, the manager of Sudbury's largest radio and television station began nightly editorials called "Stop the Strike."[35]

On the other side there were continuing signs of encouragement. Sudbury's mayor, Jim Gordon, supported by several regional mayors, condemned Inco's handling of the strike, saying in part: "The strike is driving an economic stake through the very heart of this community and for what, I ask. A stockpile? A few bucks here and there? An attempt to teach workers in this community a lesson?"[36] This was a far cry from the behaviour of the region's mayors during the 1958 strike. Encouraging words also came, ironically, from the mouth of Marsh Cooper, president of Falconbridge Nickel Mines, who stated at his company's annual meeting in April that world inventories "will drop drastically over the next four or five months and could reach danger levels by the third quarter."[37]

Workers were further heartened by the knowledge that they were preventing Inco from producing large quantities of copper and precious metals when their prices were skyrocketing. (Inco had produced 1.7 million pounds of cobalt in 1977, but the 1978–79 strike seriously cut this production while the world price of cobalt was soaring from six dollars to twenty-five dollars a pound.) By May the pressure was on Inco from several sources to produce a reasonable offer.

Compared to the offer in February of $1.75 an hour over a twenty-six-month contract, the May 6 offer valued at $3.50 an hour over thirty-six months was promising. It also withdrew the changes in the grievance procedure previously demanded. The package called for raises of fifty-one cents in the first year and ten cents in each of the next two, COLA valued at $1.65 an hour, minimum long-term disability payments of $270 a month in addition to the Canada Pension and Workmen's Compensation, and a minimum pension of $636 a month after thirty-five years of service at age fifty-five going to a maximum of $908 a month at age sixty-five after thirty-five years of service. Local 6500's negotiating committee thought it had made a major breakthrough, and indeed, so it appeared, but the membership remained militant. The three-hundred-member stewards' committee scrutinized the offer and rejected it primarily because of the pension plan, demanding instead a full pension of $750 after thirty years' service, regardless of age. The stewards were upheld by the Wives Supporting the Strike Committee. With a rallying cry of "Thirty and Out," 57 per cent of the membership voted against the contract (5,463 to 4,058). After more than seven and a half months there was still notable solidarity among the membership.

Shortly after the rejection, the two parties met again. This time the negotiating committee knew it had the membership's support, and Inco knew the offer would have to be substantial to pass. Before the month was out a new agreement was reached. It was valued at $4.07 an hour over three years (thirty cents more than the offer rejected on May 12), including the right to retire on full pension (at least $600 a month regardless of age) after thirty years, although this provision was not to be effective until June 1981. First-year wages were increased by sixty-one cents an hour (compared to fifty-one cents), but the ten-cent raise for each of the next two years remained the same; the COLA was to be retroactive to April. Although they got "Thirty and Out," the stewards' committee was still not satisfied with the value of the pension and was split on recommending the pact. The membership, nevertheless, overwhelmingly supported it on June 2 with 68 per cent voting in favour (5,983 to 2,869). The symbolic issue of "Thirty and Out" had been gained; there was no erosion of union strength; the wage gains were considerably more than Thompson had accepted. The Sudbury workers now had a base rate of $7.40 an hour for the first year (including COLA) compared with the $6.79 they had before the strike, although they had increases of only ten cents an hour for each of the next two years. The costs, however, were high. The membership had been on strike for eight and a half months – 261 days – and most were deeply in debt.

Start-up began June 5, 1979, and once on full schedule 10,900 hourly paid workers returned (about 800 fewer than when the strike began). By this time analysts were raising the possibility of nickel shortages at the end of the year while cobalt and platinum products were in very high demand.[38] Because of its duration and the number of workers out, this was the most substantial strike in Canadian history. About 2.2 million man-days were lost, equalling the annual totals for *all* strikes in Canada in the early 1970s. Symbolically it was a show of resistance at a time when the rights of many unions were being rapidly eroded by the state or threatened by capital. The issues of nationalization of Inco and workers' control of the industry were raised by the strikers and the community. Many alliances were formed between Local 6500 and the community, the wives, and the labour movement. The local became, at a time when it appeared to be doomed, a rejuvenated force, even a thorn in the side of the conservative USWA, of which it is a part. It demonstrated to Inco, and employers generally, the resilience of the working class. The tenacity of miners was proven once again.

Another test of solidarity occurred shortly after the strike's end when the strike-delayed local elections were held. Dave Patterson, the president during the strike, ran for re-election on an "autonomy" platform which called for more independence from the national and international headquarters of the USWA. Patterson won a landslide victory, gaining 4,146 votes; his closest rival had 1,019 (all opponents polling only 2,433 votes). It was the first time in twenty-one years that a president of Local 6500 had been re-elected after leading a strike (four earlier incumbents had been defeated). Patterson's disenchantment did not begin with headquarter's lack of wholehearted support for the strike. In 1977 he had actively campaigned for the militant Ed Sadlowski slate* and against the establishment candidates for the international executive.

This time the townspeople, the wives, the local union leaders, and the rank-and-file did not fall victim to divisiveness as they had during earlier strikes. They gained new energy and, most important, insight into their position within the company, the community, or the union. Only the future can tell whether this will wither or flourish. In either case, the struggle will become a symbol for future workers' resistance.

Carrying on the Battle

Workers' resistance to management in the workplace is expressed in many forms, all the way from insubordination to union-sponsored training programs. One important expression that becomes the focal point for resistance is the grievance system. It is the counterpart of the penalty system used by the company, since many grievances are lodged in response to penalties assessed against workers. As with a penalty, the higher the level of a grievance, the higher the person in the hierarchy who deals with it. Supervisors are involved in the first stage, general foremen in the second stage, superintendents in the third stage, and thereafter a grievance goes to arbitration. The system is universal throughout Inco's operations, although each union local has some differences in its contract language. Somewhat similar to the grievance procedure is the TAC system, or Total Accident Control, which has been implemented at Copper Cliff Nickel Refinery in addition to the grievance procedure. Under its provisions a worker

* For an analysis of the Sadlowski movement, see Philip Nyden, "Rank-and-File Organizations and the United Steelworkers of America," *Insurgent Sociologist* 8, no. 2-3 (1978):15-24.

can report "potentially unsafe acts" and have them dealt with as safety matters rather than grievances.

If a grievance is to succeed, the union must demonstrate that the CBA has been violated. Initially a worker brings his grievance to the attention of his supervisor. The supervisor is instructed by the company's industrial relations handbook to respond as follows:

> *If there is a doubt, give the benefit of the doubt to the company.* This statement should not surprise you. Ask any steward and his answer will be – "The benefit of doubt to the grievor (or the union)." The purpose of the grievance and arbitration procedure is the resolution of doubtful issues. However, in his area of responsibility in grievance handling, the foreman does not sit as arbitrator. He resolves those cases in which the grievor is clearly entitled to satisfaction, and, since an interpretation of the Collective Bargaining Agreement will affect the entire operation, denies those grievances in which the management is either right or the decision is borderline or doubtful.[39]

The grievance procedure tends to defuse potential conflicts and deals with them in a legalistic manner. As one senior Inco official said, "The grievances themselves are very, very healthy because so long as somebody knows that they can grieve there is a much better labour force. If they know that they can grieve anything under the sun anytime they want, then part of your problems go away, but if they can't grieve, that frustrates them. If you have always got that avenue, it is very healthy for a labour force."

The nature of grievances has changed over the years since the procedure came in with the union. A veteran miner and union official described some of the changes: "The early grievances were mostly about money, in the 1950s. That's one thing that would make workers real mad. In the 1960s you would get grievances about money, of course, but also about promotions or seniority. Monetary items again, but all aspects of it. In the late 1960s you had all of these plus a lot of safety-related issues, like refusal to work – signing out on the basis that the guy didn't figure it was safe." During the early period grievances seldom went to arbitration. From 1944 to 1960, a union official estimates, only a handful of cases were in arbitration compared with nearly a hundred cases a year since the mid 1960s. Since 1969 the number of grievances in general has risen, partly because of gains made in contract language during the long strike that year.

Grievances are designed to be the union's protection for the worker, but the union, as in many other activities, is often caught between the worker and management. It may take a case to, say, third-stage grievance and decide not to go on because it cannot win. It must then inform the worker that he does not have a real grievance (even if the complaint is legitimate it may not violate the CBA). This worker may submit only one or two grievances in his working life. To receive a letter from the union saying that he has no grievance when he knows he had a problem causes frustration, which he takes out on the union rather than the company. This capacity to turn hostility back on the union is part of the company's power. A union leader reflected on this problem:

> *The kind of legislation that came out of the war in 1945 was geared in a fashion to put the union in a bad position with the militancy of the guys. They used to solve problems either by walking out or confronting the boss as a group. This was removed completely from the process. Some unionists thought that it was a great achievement. They had this procedure. Big deal. What does it do? It takes away the collectiveness by which the guys dealt with problems in the past, and it comes down to one guy with a problem for this violation of the contract.*

Most workers never file grievances; it is estimated that only 5 to 10 per cent have ever filed, although a higher proportion may have had cause. One surface worker told us why he has never put in a grievance:

> *I have never filed a grievance since I've been here* [twenty-three years]. *Never. I've had cause to a few times, but I have always gone to the boss. . . . I should have put a grievance in about my lousy converter. You get an old*

converter and you can't do too much with it. I couldn't get it hot enough. It was leaking too much, so I couldn't get it hot enough for separation. It sat there for about fifteen minutes and the bottom burnt out. One of the bosses was standing there, and I told him what I had in the pot and said, "There's a crane up there. Tell him to pick it up and get rid of it." He said, "Oh, that's all right. That's good." He walked away and it happened right after. It burned out. If he had told the crane man, "Pick up that pot; he's out of matte," then they could have dumped it right away and that's it, but he didn't do it. . . . I got a warning, a Step Two. The shift boss came up and asked me what happened, but I didn't want to involve the other guy, the other boss. He just got his job, so I didn't want to get him into trouble.

Other Inco workers have filed grievances and lost them. Whatever happens, they feel that the audacity is held against them by the company. As one young miner put it, "If you win or not, it's still on your record, so why bother?" He had filed two grievances for what he felt were just causes (once because a junior person was sent to driller school before he was and once concerning overtime) and lost both. He says of the union, "They haven't helped me," but is quick to add, "I'm glad it's there, compared to what I've heard from the guys who worked here before the union." (For many miners the union is like an insurance policy; they hope they will not have to use it, but they are happy it's there if they really need it).

Union stewards say that management can and frequently does turn the grievance process against the union. It is true that stewards encourage the filing of grievances to demonstrate dissatisfaction with management, but management often counters with more rigid exercises of power. For example, permission to take leave is normally granted quite freely, but if a steward seems to be pushing grievances in his work area, the shift boss can begin to withhold leaves from the men and make the steward the scapegoat. When safety and health issues are at stake this can be a serious matter.

Much of the politics between the company and union revolves around grievances. Each of the three major Inco locations has a quite different experience with grievances. Fewer grievances in proportion to numbers are filed in Thompson and Port Colborne than in Sudbury. In Thompson the high turnover is partly responsible; the transient workers often do not know their rights or do not care, putting up with adverse conditions because they know they will soon be moving on. A related factor is the two-month probationary period, during which workers pay union dues but lack union protection. A greater proportion of the workforce in Thompson than elsewhere is affected by this arrangement. In Port Colborne fewer grievances come up because of the wariness of workers. Even though 6200 was once considered one of the union's most militant locals, today it is quite inactive. There have been only eighteen arbitration cases, for example, since 1969, and the company won fifteen of these. The two grievance commissioners (whose decisions do not set precedents) have been more lenient, one giving the company and union each about half the decisions and the other tending to decide in favour of the company.

The grievance system is very expensive. If the company and union cannot come to an agreement on the grievance, it can be taken to arbitration. Here the costs mount, averaging about a thousand dollars a case. In Ontario the waiting period before an arbitration hearing takes place may be four to six months. Even though the company and the union split the costs of arbitration, the union feels the financial pinch more, and in some locals arbitration costs are becoming prohibitive.

A study of the penalty system gives some sense of what goes on in the workplace. The information in Table 25, gathered from Local 6500, shows why workers were penalized for misconduct (these figures apply only to penalties that were grieved and went to arbitration). The most frequent penalties are for absenteeism. During the period from 1975 to 1977, which these statistics cover, 232 warnings for AWOL cases were grieved (these cases seldom go to arbitration, being resolved or dropped at a lower level). The company won 57 per cent of the cases while the union was successful in 43 per cent. The other most common reasons for penalties are unsafe practices, failure to follow instructions, and refusal to work. All are indications of some form of alienation. The more severe

Table 25

Warnings for Misconduct Penalties Grieved at Inco's Sudbury Operations, 1975 to 1977
(1-in-10 systematic sample)

Reason for penalty	Union win	Company win
Loafing	3	10
Horseplay	5	8
Leaving early	8	16
Unsatisfactory conduct	3	10
Unsafe practices	8	17
Sabotage	2	1
Lateness	1	6
Insubordination	3	8
Failure to follow instruction	7	14
Refusal to work	1	26
Violation of company standards	3	6
Absent from work area	1	9
Disobeying orders	–	3
Failure to do work assigned	2	1
October 14th Day of Protest	6	3
Unsatisfactory work	2	5
Drunkenness	–	2
Negligence	1	1
	56 (28%)	146 (72%)

manifestations of alienation, such as verbal abuse or threatening a supervisor (often called unsatisfactory conduct), are likely to be punished by Step Four or discharges. From the 1975 contract to the summer of 1977, about 14,000 grievances were filed. Table 25 shows a sample of grievances involving penalties that were fought by the union. In the cases reviewed, the company was notably more successful than the union. During the same period (1975–77), 143 cases were taken to arbitration in the Sudbury operations. Of these, the union won 22 per cent, the company 64 per cent, and 10 per cent were compromised. Wages, overtime, and other monetary disputes were most common (seventy-nine), followed by insubordination, refusal to work, and swearing at or abusing a supervisor (eighteen), unsafe practices (fourteen), denied work by the company (twelve), horseplay and gross misconduct (eight), loafing and leaving early (seven), and absenteeism (five).

Maintenance workers at Inco are the most militant. The consensus among union leaders is that people with higher skills are quicker to fight for their rights than others. They also have had more union experience than most workers. In the mines there is also a difference. Maintenance workers in the mines are not on bonus and are therefore not involved in the drive for production that sometimes causes other hourly workers to overlook grievances. Moreover, they tend to be more conscious of safety than bonus workers. Most of all, maintenance workers are conscious of preserving their trade and resist encroachments on their domain. Frequently they appeal to the grievance procedure to ensure this protection.

Grievances are important not only for the individual but also for the shop steward who files the grievance, for the reputation of the union and its ability to protect the worker, for setting precedents, and for the accumulation of issues to bring up during contract negotiations. They are part of the politics of the workplace.

Unlike workers in some industries, miners seldom engage in another form of resistance - sabotage. The danger of the work discourages this kind of action. Miners know that any damage is much more likely to affect themselves or their fellow workers than the company. The work on surface is hazardous also. There are, of course, a few incidents where a worker will purposely break a machine or damage some company property, but there is little evidence that this practice is widespread. One veteran miner in Sudbury was asked about sabotage and responded, "I don't see it. I've never seen it. The only time I've ever seen anything like that was in 1966, when we had trouble with the strike. Other than that, I haven't seen any reactions of that sort." A much more common expression of dissatisfaction among mine workers is absenteeism.

The rate of absenteeism at Inco runs between 8 and 10 per cent compared with the national rate of 4 to 4½ per cent. Absenteeism includes both being sick and being AWOL. Because of absenteeism the company must overstaff, employing an extra hundred workers for every thousand it needs for production. Younger people tend to go AWOL while older people, particularly those over fifty, are likely to be absent for health reasons. Many miners work an unofficial four-and-a-half-day week, leaving on Friday afternoons or early on a weekend shift, giving "sickness" as their excuse.

It is difficult to say how much absenteeism results from health reasons and how much from simply "blowing a shift." Both reasons are common. Absenteeism expresses a worker's dissatisfaction with his work and is a way of thumbing his nose at the company. Many miners recount their frequent absences from work with great pride as an expression of their freedom.

Others go further and quit, saying that they "don't have to take it any more." This form of resistance is seen in turnovers in the labour force. These vary with the location. At Port Colborne there are virtually no turnovers; at Sudbury the turnover rate has been relatively low in recent years because of cutbacks; but in Thompson the rate is so high that attrition has taken care of many who would otherwise have been laid off. In 1973 the Mining Association of Canada commissioned a study of turnovers in the industry that provides some insights into the characteristics of turnovers, as Table 26 illustrates.[40]

Table 26

Turnovers in Canadian Mining by Occupation, 1973

	Per cent turnover
Unskilled labour	127.8
Skilled miners	49.8
Office workers	41.7
Technical staff	27.6
Management staff	22.2
Supervisory staff	22.0

SOURCE: *Canadian Mining Journal* 96, no. 5 (1975): 26.

In absolute terms, in the Canadian mining industry some 19,500 unskilled labourers and 21,378 skilled miners left their jobs during 1973. The rates drop noticeably for staff, but even office workers approach the turnover rate for skilled miners. This may well be because some are women employed at clerical work for the company, and when their husbands quit their jobs, the women are forced to follow. Miners who quit seldom remain in the same community because the company they have left is usually the only major employer. Turnovers and the inability to replace workers as fast as they quit create an average shortage of 6.4 per cent of the unskilled and 4.7 per cent of the skilled mining labour force.

Mining communities are usually divided into two populations, "a stable group which includes the married couples with children who have lived in the town for some years and have roots there, and a group of young, single male workers who have no ties with the community, do not own property and do not wish to do so."[41] The turnover rate among single mining employees in 1973 was 125 per cent; for married men it was 34 per cent. Other factors that contribute to the turnover rate are the size of the mine and the distance from a major city. Mines employing under a hundred or over a thousand workers tend to have the lowest turnovers; the ones between have substantially higher rates, especially those with two hundred to five hundred workers. The turnover rates are lowest in mines close to major cities and highest in mines fifty to a hundred miles from a major city, although those farther away also have high rates.

Labour shortages are a serious problem for mining companies because without labour, especially skilled labour, they cannot produce. They have reacted to these conditions by moving into heavy capital expenditures to reduce the amount of labour required. As the *Canadian Mining Journal* has said, "Many young people are unwilling to commit themselves to life in a mining community and work in a mine. One solution to this problem would appear to be new technology to reduce the labour content and also improve the working environment for miners."[42]

At Thompson, where Inco encounters its main turnover problems, a "stable" labour force of about 40 per cent of the population stays a year or longer, but the rest turns over at a very rapid rate. Table 27 outlines the history of Thompson's turnover record. Although the turnover rates vary, they have been consistently high. The 1977 rate was the lowest, but it was also the period when attrition was being used to reduce the labour force, so that few people were being hired. Otherwise the turnover has generally been equal to or greater than the size of the total labour force.

* * * * *

A Thompson worker who arrived in 1962:
In those days they never used to call you by your name;

Table 27

Inco Labour Force Annual Turnover in Thompson, 1960 to 1977

Year	Per cent	Year	Per cent
1960	223	1969	184
1961	217	1970	125
1962	186	1971	110
1963	122	1972	100
1964	127	1973	84
1965	97	1974	130
1966	118	1975	126
1967	163	1976	104
1968	180	1977	66

just by your number. My number was —— and somebody else might be, say, 1207. They would just say your number. Your foreman would come in the morning, and he just had a list of numbers. But that was too hard because people were turning over so often. Very few people stayed very long. I was unusual in that I stayed. After a year or so he got to call me by my name.

People would come from many lands. A lot of people who used to come here were running from something else. A lot of them jumped ship. I remember quite a few guys that worked with us; they would jump ship in Churchill and come to Thompson, get a job under an assumed name, and the manager would come along and pick them up. Once in a while they were rounded up, and I'm sure Inco used to supply the hiring list to the Mounties so they could check it out, because there were a lot of criminals.

* * * * *

The situation in Thompson is attributable to the workers' reactions against both working conditions and living conditions. The high cost of housing, heat, food, gasoline, and clothing eats away at their incomes. The lure of the "big bonus" turns out to be an illusion for those without mining experience. They are isolated from major urban centres. All these factors contribute to the turnover, but so does the work. Most people simply are not willing to put up with the danger and physical effort associated with mining. They try it for a short while but then tell the company they have had enough. Like absenteeism, labour turnover has an impact on the company, but as a form of resistance it is not very effective unless the supply of labour runs out. Between 1960 and 1975 about 45,000 men worked for Inco in Thompson when the total labour force was only about 3,000. Most of these were young men between the ages of eighteen and twenty-five. This is also the age group with the highest unemployment rate in the country. Unless unemployment elsewhere drops very low this labour pool will continue to exist.

A final form of workers' resistance is the miner-as-a-trade program. The purpose is to certify mining apprentices as the plumbing, mechanical, and other trades do that require "tickets" to practise. Theoretically it is the workers' strategy for coping with the specialization and fragmentation of work characteristic of mechanized mining and modular training. In practice it is not able to meet this challenge.

Inco can simply install its MTS program, but the union has to convince the province and the province persuade the mining companies to implement the miner-as-a-trade apprenticeship. Although miners have been certified in several European countries since 1951, the first program in North America began only in January, 1975, in the province of Manitoba. The apprenticeship covers a three-year period and requires eight weeks a year in school in addition to the time spent working in specific areas. The program is provincially controlled, and all mining companies participate, however reluctantly. The NDP government, which implemented the program, did not make apprenticeship mandatory, as it is in most trades. The province also gave "grandfather" tickets to about three hundred miners with four or more years' experience in 1975.

Mike Geravelis from Local 6166 in Thompson outlined the apprenticeship program for the Ham Commission:

> The apprenticeship program is a three-year program and as you can see the mine helper, the true requisite for apprenticeship. . .has to be working underground for three months, and in that three months. . .you will notice the assigned duties, [learning] scaling procedures and ground control, emergency procedures, layout of the mines, escape

> routes, care and use of personal protective equipment, cage signals and operation, check in and out procedures, basic safety rules and regulations and personal responsibilities as per the Mining Act. . . . After he has served his three months initial training. . .he goes into [service and utilities] for a total of 47 weeks. In that 47 weeks he in turn rotates through 11 classifications; for example, motorman, switchman, cagetender, shaft serviceman, skiptender, crusherman, topman, sandfill man, slusher and scraper operator, miscellaneous services, underground equipment preventative maintenance and mine rescue and first aid.[43]

Miner-as-a-trade in its present form is not going to revolutionize the mining industry: at present it is a mere drop in the bucket. In 1977 the first graduates completed the course – six of them. At present only about twenty people are enrolled in apprenticeship in Inco's Thompson operation. This is not to say that the program could not be important, but as long as it is not a prerequisite to being a miner, it is unlikely to countervail the tendencies of detail labour inherent in mechanization and MTS as Inco is implementing them.

There are several reasons why apprenticeship has not been attractive to Thompson miners, aside from the fact that one can get into mining without it. At present one can still attain top wage rates in mining without certification. During the apprenticeship, individuals make lower rates and obviously forgo bonus. Then too the program has not been recognized outside Manitoba, and many miners, particularly in Manitoba, are unwilling to commit themselves for three years. And the miner-as-a-trade, unlike the traditional electricians, mechanics, carpenters, welders, and plumbers, is tied to one industry, which is notoriously unstable as an employer.

Inco officially supports the program in Manitoba because it is required to by the Manitoba government, but it is clearly not enthusiastic. The Ontario Mining Association is attempting to prevent a miner-as-a-trade program from being introduced in its province, although it will probably accept a five- to seven-week training plan as a result of pressure from the unions and the Ham Commission. The company obviously dislikes the program because miner-as-a-trade attempts to acquaint the worker with all aspects of the industry while the industry is moving toward more specialized detailed work. As a senior Inco official argued, miner-as-a-trade "tends to be very broad. It covers all service work, all production work, all development work; very seldom can an individual use all of those because he is either in production or in development. If he is producing muck in a stope, he's not sinking a shaft; if he's driving a drift, he cannot be operating on service. Physical location precludes doing all these jobs."

In the following comments, another senior Inco manager gave some indication of the strength of management's feeling against the program:

> *Thompson miners maintained that if they didn't go that route they were going to go to the legislature and [Inco] decided that they would work within that structure to get it put together; but it isn't working. In my opinion it is not working in Manitoba. They have only got something like twenty-five apprentices in the whole system of four thousand miners in the province, and it is not really meeting the objectives that they set up. I think there are much better ways. It just gets too large for the people to really effectively utilize. They can learn, but they can't apply. They can't use what they are learning effectively in the workforce because they are learning more than is required in the workforce. They are teaching them all the different development mining methods, all the different production mining methods, and then all of the service – the shaft, the shaft sinking. They are talking a three-year apprenticeship.*
>
> *You are then a tradesman, but how do you effectively use the skill that you've gained? That's a problem. The electrician, he goes and does the work that is required to be done. Now some of that is very high skill and some of it is very low skill, depending on that particular area. These fellows can bid on particular work settings and they are only going to use the skills necessary to do the work in that area. They have all of this other skill that they have acquired through the miner-as-a-trade, but they are not applying it at all and won't apply it. Now, if you want to*

change all of the seniority provisions of the collective agreement, that maybe starts to make a lot more sense, but there is no way they are going to do that. That's heresy.

Unless there is a dramatic change of government in Ontario or an even more dramatic change in the social organization of work at Inco, the miner-as-a-trade program is doomed to be window dressing that can be pointed to on occasion to satisfy union complaints. Unless it becomes compulsory training before anyone can work in the mines, it will not be utilized extensively by miners. And, of course, mining companies prefer their own training schemes, which they can control and dovetail with the implementation of new forms of technology.

Without doubt the unionization of hourly rated Inco workers has improved their working conditions. It has removed *some* of the arbitrary power of capital in the workplace, although it has little effect on such important actions as massive layoffs. It has improved wages, although within limits; the union has only been effective in bargaining when it has had the leverage of high demand for nickel during government stockpiling or wartime. Unions have not, however, been successful in reducing injuries or deaths at work. They have limited ability to combat the power of capital. They can, and do, collectively articulate many of the demands of workers. They protect individuals to the extent possible, but they cannot unilaterally determine the conditions or terms of employment. The company can still fire workers, although usually only after proper action.

Trade union leaders continually find themselves in contradictory positions. To acquire legitimacy they have to accept the rules and procedures of capitalism, which means accommodating to capitalism. In this the state has an important role in controlling union actions. The leaders are under pressure from the membership to improve working conditions, ensure employment, and maintain wages, yet capital, not the unions, has control of these matters. Unions thus serve capital by deflecting some hostility away from it. Although unions provide educational opportunities and at times tend to draw workers together in solidarity, they cannot always control the actions of workers. When union leaders are unable to contain these, the leaders themselves come under threat from the companies or the state. At one and the same time union leaders must mobilize workers' resistance if their actions are to be effective, yet they must contain these actions to maintain their own legitimacy in the eyes of the state.

CHAPTER TEN

A Political Economy of Mining

Work has been fundamentally reshaped in Canadian mining in the post-Second World War era. Capitalization by the companies has been expressed as mechanization underground and automation on surface. Except among maintenance workers (who are currently threatened by modular training in their status as tradesmen), the effect has been to decrease traditional skills in the mining industry. Skilled work has been shifted from the workplace to science-intensive research and development centres and to firms engaged in machine manufacture. Inco's two main laboratories for research in mineral manufacturing applications are located outside Canada, one in the United Kingdom and the other in the United States; only process research is done in this country. Moreover, the lion's share of machinery used in Canadian mining's capitalization program is conceived, designed, and constructed outside Canada. All this means that Canadian workers are left with the jobs calling for the lowest skills (and these are growing fewer with greater capitalization) while workers outside the country are given the technically advanced work in the fields that tend to be expanding. The picture is not unique to the mining industry. Some time ago the Science Council of Canada warned us of this general pattern[1] and has recently documented its effects.[2]

We have examined the ways in which capitalization has decreased the number of jobs and changed the available jobs to eliminate craft skills, thereby affecting the workers' capacity to combat the pressures on them.* Management's strategies of internationalization and diversification have also been reviewed. It is now time to take a somewhat wider view of mining and discuss the structure of research, the mine supply industry, and the nature of the relationship between the Canadian state and mining companies, and finally to make a tentative extrapolation of some implications of developments in the class relations within mining for the broader class structure.

Buying Research and Technology Abroad

Whereas the era of the independent commodity producer in Canadian mining was relatively brief because of capitalist penetration, reliance upon the skills of miners in hard-rock mining during capitalist production has been relatively long. The actual labour process had responded with resilience to mechanization until the late sixties, as had most surface operations to automation. As a result, mining remained, until relatively recently, a labour-intensive industry. The cost of labour and the skill requirement of the labour force, however, have been higher than in many labour-intensive industries. The processes outlined in the earlier chapters are changing this condition.

During the rapid capitalization of the past decade, one would have anticipated a backward linkage from the mining industry to greater research and development in Canada. In fact, the opposite has happened. While other industrial nations have been expanding their proportion of employment in research and development at the rate of 8 per cent a year, in

* This is not to say that the miners' solidarity has been decreased by the homogenizing effects of capitalization. Indeed, the long 1978–79 strike at Sudbury may suggest that the remaining workers will behave more cohesively as a result of the levelling of skills and threats to their jobs. Deskilling weakens the individual power of tradesmen and craftsmen, allowing management to replace them more readily. At the same time, however, it weakens traditional fractions within labour based on skill levels. Until now unions have been unsuccessful in preventing the decline in the numbers of workers resulting from greater capitalization. The reduction means, of course, a smaller union membership, hence fewer union resources.

Canada such employment has actually fallen.[3] In Canada, 95 per cent of patents are registered to non-residents, and two thirds of them are owned by residents of the United States. The mining supply industry has some responsibility here. Writing in 1941, E. S. Moore recognized that there was a fundamental flaw in Canada's resource development: "Control over Canadian mines by Americans and the extent of American investment have involved substantial imports of capital equipment and the establishment of American branch plants."[4] Since then the problems have multiplied.

When mining had low capitalization (that is, little machinery), most of the value in *extraction* stayed in Canada because the work was performed by Canadian workers using labour-intensive techniques of production. With high capitalization, however, much of the value is added through the use of machinery operated by fewer workers. The value transferred to the ore through the machinery is the stored-up value contained in the machinery. Since much of the machinery is built outside Canada, that value is not Canadian. Not only are there fewer jobs for Canadians but also the jobs that are lost are in the most scientifically advanced occupations. The labour that goes into building machinery is consumed as that machinery is used. This stored-up labour is imported along with the machinery built outside Canada. When imported capital-intensive equipment displaces the labour-intensive activities of Canadian workers, the net effect is to decrease employment in Canada.* The major benefits derived from capitalization in mining are therefore exported. As John Britton and James Gilmour have noted:

> The forest and mining industries in Canada comprise a large market for engineering and machinery. In contrast to other industrial countries with similar activities, such as Sweden, Canada has incurred heavy net deficits in the engineering and machinery requirements of these industries. Explanation for the underdevelopment of these in-

* This argument is derived from the proposition that "machinery, like every other component of constant capital, creates no new value, but yields its own value to the product that it serves to beget" (Karl Marx, *Capital*, 1867; reprint ed., New York: International Publishers, 1967, vol. 1, p. 387).

dustries lies in the equipment purchasing patterns of foreign-owned forest and mining firms by-passing Canadian engineering, consulting, and machinery manufacturers.[5]

It is therefore important to place mining workers within a larger class context. They are workers with only their labour power to sell, and the demand for that labour power is being decreased *within* the mining industry. Moreover, in the broader labour force there is less opportunity to sell their labour power because the labour contained in the machinery that is replacing them in mining is being added outside Canada. The overall effect, of course, undermines the value of labour but strengthens the power of capital to extract surplus value. These tendencies lead to high unemployment and high profits, a condition prevalent throughout Canada's recent history.

Inco has developed little extraction technology; instead, the company relies on the mining supply industry and co-operation with that industry. In 1974 the *Canadian Mining Journal*, certainly not a constant critic of the mining industry, published a severe comment on the state of research and development in Canadian mining technology. It said, "The equipment supply industry in Canada at the present time is dominated by the presence of foreign-owned suppliers. The strategy of these companies appears to be one of importing new foreign technological advance as it becomes available, thus few funds are allocated in Canada by these companies for research and development of new products to serve mining. . . . In fact, there is evidence that innovation by mining companies in Canada is on occasion transferred to the foreign parent company of a Canadian supply operation. The parent company exports the innovation to the Canadian mine."[6] These observations were supported in a recent study by P. R. Richards and his associates, using a sample of thirty-two mining equipment companies. They confirm the claim that the supply companies operate as branch plants and are set up to get access to the Canadian market; they have no interest in research and development in Canada. Moreover, the study establishes that "mining firms do not manufacture their own process equipment, but rely on suppliers."[7] This means, of course, fewer production jobs in Canada, but it also means fewer jobs in research and development. At least on this point there appears to be a basis for common interest between production workers and technical workers.

* * * * *

A senior Inco engineer, questioned about Inco research on equipment and technology:
Well, basically we have dealt with manufacturers and tried to get some of the supply manufacturers that produce different types of equipment to deal with this. A lot of the problem is that a great part of the mining equipment is really custom built. It is no mass production line. There are a few things that we can adapt, for instance the Caterpillar front-end loader. We can manage to fit something like that in a production line. The rest is rather expensive.

* * * * *

There have been some developments recently in Canadian research, but the impact appears slight. For example, six "Canadian" mining companies (Falconbridge, Inco, International Minerals and Chemical Corporation, Noranda Mines, Rio Algom, and Sherritt Gordon) have founded the Mining Industry Research Organization of Canada; its program for 1976 included "the development and testing of an improved lighting system, more effective noise protection, and wearable air filtering equipment for use when needed – all involving redesign of the employee's protective helmet."[8] It does not appear that the organization will be involved in developing extraction machinery or in encouraging manufacturing of this equipment in Canada. The mining companies have also entered into a cost-sharing program with the federal government called Development and Demonstration of Pollution Abatement Technology, but again, this is "to assist the mining industry in improving working conditions in the face of difficulties arising from deeper workings and the more intensive use of machines."[9] This research may help the mining companies adapt the miners to the dangers associated with the new technology, but it appears to have little to do with the development of technology that would improve extraction techniques themselves or foster establishment of a Canadian mining supply industry that would expand Canadian employment.

Help from the State

During the early stages of mining in Canada when capitalist relations of production were penetrating petty commodity forms, the state had an active role, particularly in changing regulations governing the rights of access to property. Governments expected in this way to develop a lucrative tax base through capitalist mining. Since then a close relationship has continued between mining companies and the provincial governments in particular, mineral rights being in the domain of the provinces (with the exception of uranium).

From the outset the mining companies tended to perform only a minimum amount of processing in Canada. At various times governments have applied pressure to increase the amount of Canadian processing, like that leading to the construction of Port Colborne Nickel Refinery. But the means of pressure have most often been financial inducements to the companies and concessions that further improve their profitability. None of these government pressures have been effective in developing industries to use the raw materials produced, most of which are destined for export markets. In 1943 the Ontario government appointed a royal commission to inquire into a strategy to strengthen the province's mining industry. As Ross Hynes has observed in his review of provincial mining policies, "the Commission reported largely, if not exclusively, on the basis of testimony given by industry representatives." He goes on to say,

> The report of the Royal Commission recommended the total reform of the provincial securities legislation, the expansion and modernization of the government's education and geological information programs, the establishment of a construction program for mining access roads, and the intensification of provincial government pressure on the federal government to reduce its taxes on the mining in-

dustry . . . the government's mineral resource policies in the post-war years would appear to have been derived largely from the Commission's report.[10]

Millions of dollars were spent by the Ontario provincial government, often in conjuncture with the federal government, on programs such as "Roads to Resources" as inducements to further mine development. These programs were supplemented with publicly financed geological surveys. But the most significant form of inducement was tax breaks. After the Second World War the governments of the mining provinces made vast concessions in the form of royalties, tax holidays, and depletion allowances, along with tax concessions from the federal government. Hynes found only two tax adjustments for mining companies in Ontario between 1945 and 1968: "In 1946 the government of Canada allowed provincial mining taxes as a deductible expense for federal income tax purposes. . . . In 1957, the mining industry, thriving in a period of general stagnation, was seen as the logical candidate to meet the government's revenue needs, and taxes were again raised." These were accompanied, however, by other actions counteracting the effects of the adjustments:

> In 1947, the deductibility of the mining tax for provincial income tax purposes was established, a move previously seen as unnecessary because the rates had been so low. More significant was the exemption from the mining tax of 50 per cent of the profits of a mine in the first year of operations. This was followed in 1948 by the recognition of exploration expenses. A depletion allowance of 33 1/3 per cent of mining production profits for purposes of computing corporate taxable income was added in 1957.
>
> The 1957 tax rate increase was made more palatable for the industry by the introduction of a three-year income tax holiday for new mines, replacing the less generous exemption from the mining tax of half the profits of such a mine.

There appeared to have been a change in the nature of taxation at the beginning of the 1970s with the recommendations of the Carter Commission, Hynes continues. Ontario, in response to federal policies, "introduced a higher, steeply graduated rate structure for its mining tax."[11] But by the end of the decade some provincial governments were in retreat. Not only was Ontario granting generous allowances of 30 per cent for refining and 35 per cent for processing but it was even granting tax concessions for the refining of nickel outside Canada. Ontario Treasurer Darcy McKeough in 1978 "promised that Inco Ltd. could deduct from taxable profits the costs of its plant in Wales and that Falconbridge could deduct the costs of its plant in Norway." He is reported to have said that the provincial government was "not giving up the dream of more processing in Canada. We're being more realistic about what's going on and removing a disincentive to more mining and explorations."[12] Inco's strategy of internationalization has an effect on more than the workers. Indeed, the Ontario Ministry of Natural Resources learned the lesson well, arguing in a report that if Inco's Indonesian and Guatemalan operations prove too profitable, "production cuts will continue in Canada rather than overseas and the situation would be aggravated by any further increases in taxes in Canada, or by any other measures that lower after-tax rates of return."[13] The leverage of multinationals is indeed powerful.

In Manitoba, the New Democratic Party government proved less amenable than the Progressive Conservative government of Ontario. It had begun a program of mineral exploration in 1974 when it required an equal public partnership in all exploration conducted by private mining companies. Inco and the other large mining companies of the province opposed the NDP government and its policies. With the election of Sterling Lyon's Progressive Conservative government in October, 1977, their efforts paid high dividends; the new Mines minister, Brian Ransom, suspended the program. His deputy, Paul Jarvis, gave further solace when he said that the government "assured the mining industry that adjustments will be made to the existing mineral resource taxation regime of the province . . .that will improve the mining investment climate." He added, "Exploration and mining should be left to the private sector." The Lyon government fulfilled its promise, reducing mining taxes, effective January 1979, to a flat rate of 18 per cent, replacing a system put in place by the NDP of 15 per cent for less than 18 per cent return on investment and 35 per cent on income over that amount. To sweeten the deal further, a five-year freeze was placed on the hydro-electric power rate for the mining companies.[14]

Provincial governments have other ways to subsidize mining companies. The nationalization of the Northern Ontario Power Company in Ontario in 1945, for example, led to major savings. Hynes says that "following the purchase, [Ontario] Hydro renegotiated its rates with the mining companies, and arrived at a 25 per cent cut. Rates charged to the communities were left untouched."[15] We have already seen the major concessions on electricity rates made when the Thompson, Manitoba, facilities were built. These savings through provincial agencies are particularly important for Inco; as one of their senior Sudbury engineers said, "We produce, presently, about 35 per cent of our power, and for the rest, we are the biggest single customer of Ontario Hydro. Inco buys more power than any other single company in the province from Hydro."

Throughout its history, the mining industry in Canada has been favoured by state policies. From the perspective of governments, the mining companies are crucial to the health of the economy. When they fail to produce, everything else slows down. Mine products, as discussed in some detail in Chapter 4, are critical to the nation's balance of payments. Moreover, "minerals account for 55% of all freight handled by Canadian railways and 60% of all cargo loaded aboard vessels at Canadian ports."[16] But Canadian state leaders have been extremely shortsighted in their approach to Canada's non-renewable mineral resources. They have presided over their development as raw materials for the industrial capacities of other nations but have done little to foster an industrial capacity in Canada.* Canada's distinction of standing third in world production of minerals is only overshadowed by being first in mineral exports. With those exports go jobs; manufactured

* Britton and Gilmour note that "Canada has little or no protection on a wide assortment of machinery used by resource and resource-processing industries, e.g. agriculture, mining, oil, and gas. In 1976, 67 per cent of all imports (by value) of drilling, excavating, mining, oil, and gas machinery entered Canada duty free" (*The Weakest Link: A Technological Perspective on Canadian Industrial Underdevelopment*, Background Study No. 43, Ottawa: Science Council of Canada, 1978, p. 51).

products are imported again for the Canadian market, leaving the valuable jobs outside the country on an even grander scale than occurs simply within the mining industry *per se* (as discussed in the first section of this chapter).

Class and Mining

Earlier we explored two aspects of class transformation for Canadian mining. The first was the transformation from petty commodity production to the capitalist mode of production. We saw that the autonomous organization of work, craft skills, and the bonus system, all characteristics of this earlier mode, were not completely destroyed in the initial transformation to capitalism. Yet these remnants are threatened by the second, more current, aspect. The most obvious change in the property relations of mining occurred when the petty commodity mode was destroyed by the capitalist mode, but there have also been significant changes within capitalism itself. Mechanization is obviously the threat to the miner's traditional position in the industry.

There is an initial lag between the material development of the forces of production (such as the introduction of new equipment and techniques) and the social relations of production. For instance, different qualities of labour are required for operation and maintenance of a traditional electrolytic refinery on the one hand and a carbonyl process refinery on the other. This is not a mere mechanical relationship. Capitalists are seeking to minimize their variable costs in the form of labour while making the most of their fixed costs in equipment, thus reducing the amount of labour required. On the other side, labour is fighting against its own elimination and the changes in the capitalists' demands. This was evident at the Copper Cliff Nickel Refinery, where the new "rules" issued by Inco concerning operator versus maintenance tasks brought a flood of grievances. In this case the company attempted to minimize the reaction of labour and maximize its own control through extensive training programs and application of human relations techniques. Thus there is a dynamic operating between the forces and relations of production, each of which has implications for the other. The fundamental relation is social, whereby capital controls labour in order to maximize profitability and uses the technical division of labour as a means to accomplish this end.

In the mining industry, at least, the major changes in the relations of production take place primarily in new facilities or in drastically revamped operations, thus minimizing direct conflict with the workers currently performing these jobs. For example, ramp mining has been established in new mines (Levack West) or in mines that can no longer be worked by old techniques (Creighton No. 9); scooptrams were introduced into traditional stopes as captured equipment while the same organization of work continued, so there was little resistance from the miners. Similarly, automated milling required a new facility, as did automated refining; in the older settings only minor adjustments were made. The implications for the company as a whole are obvious, but workers engaged in the immediate work process are not directly affected. Inco seldom lays people off to coincide with internal technological change; it tends to schedule its large-scale layoffs to occur prior to contract negotiations or along with closing down major operations, as in Port Colborne. This strategy has lessened the workers' effective opposition to new techniques. Overwhelmingly, technology has been used by management as a weapon in the class war. It is used both to decrease the labour force requirements and to reduce the autonomy that workers have derived from their skills and their control over the production process. All too often workers have not even been aware of the broader implications of this strategy or were too powerless to resist it effectively.

With capitalization has come change in management strategies towards the workers. Underground, management tries to move away from the traditional responsible autonomy of mining crews and toward greater direct control. On surface there has always been more direct control of workers (with the exception of such craftsmen as skimmers or strippers, who have been on bonus), but when automation is introduced, new strategies are devised. The automated plants, Clarabelle Mill and particularly the Copper Cliff Nickel Refinery, require a different kind of labour from that in the older labour-intensive processes. Workers perform patrol and maintenance duties rather than detailed labour. Management appears to be encouraging responsible autonomy, but in fact, workers and supervisors find that they are to be held accountable without making meaningful decisions. There is a great deal of responsibility and very little autonomy. They have virtually no control over the actual work process, which has been programmed into the equipment.

Trist and Bamforth have examined some of the social and particularly psychological implications of changing methods of coal mining from the traditional "hand-got" system, where miners worked in pairs performing all the tasks in the mine with the assistance of a trammer to move the coal. This method had been essentially a craft in which miners often worked on contract with the colliery managers. "Longwall" mining destroyed this system of work relations by introducing mechanization and a new form of organization. A larger work group in which the members had specific tasks and there was a more hierarchical authority structure was called for. The new system meant "a greater intolerance of unsatisfying or difficult working conditions, or systems of organization, among miners, even though they may not always be clear as to the exact nature of the resentment or hostility which they often appear to feel."[17] Workers in Canada's nickel mines also express the same kind of resentment, sometimes through legal and wildcat strikes but frequently simply by leaving the industry. The workforce at Thompson, for instance, is built upon a floating reserve army of young male workers with low skills. Workers are often of two minds about new technology. They see that it reduces the amount of bull-work, and this is a great relief. On the other hand, it destroys traditional work relations among experienced miners, and they resent the change. This is not simply resistance to something new. The resentment is directed toward the loss of autonomy and skill. On some extreme dramatic occasions, such as the 1978 layoffs in Ontario, they come to recognize the relationship between capitalization and permanent loss of jobs, but feel powerless to do much about it.

The main impact of mechanization is to decrease drastically the need for both skilled *and* unskilled labour. The workers are replaced by what are euphemistically called "semi-skilled" labour or machine operators. In other words, both heavy manual labour *and* craft skill give way under capitalization to workers who tend equipment with built-in work processes. This does not happen automatically; labour resists management's strategies because many jobs are lost and the best-organized fraction of the working class, the tradesmen, are often directly threatened. As Katherine Stone so eloquently put it, "the prize in the class struggle was control over the production process and the distribution of the benefits of technology."[18] The workers' resentment is illustrated by the grievances at Copper Cliff Nickel Refinery, but at least in this instance the company succeeded in expanding the tasks of the operators without increasing wages. The struggle continues as Inco imposes modular training that could well expand the responsibilities of the operators, and negotiations are taking place with the union over co-operative means for determining wages.

Katherine Stone pointed out that "technology, by itself, did not create today's labour system. Technology merely defined the real possibilities."[19] Class struggle focuses on control over the production process and the distribution of the expanding surplus technology makes possible. Having broken the power of craftsmen and eliminated most labourers, capital can afford to increase the wages of "semi-skilled" workers and still appropriate the lion's share of the surplus. It does open, however, the possibility for broad-based action by workers to appropriate the means of production.

Throughout, it has been argued that it is not the introduction of technology *per se* or the technical division of labour that has caused negative effects but rather the social relations of production and the way technology is used as a strategy by management to minimize the resistance of workers. As Marx observed of the initial industrial revolution, "It took both time and experience before the workpeople learnt to distinguish between machinery and its employment by capital, and to direct their attacks, not against the material instruments of production, but against the mode in which they are used."[20] It is no longer possible (or even desirable) to return to petty commodity production in mining. The forces of production have become "socialized" by giant multinational corporations. The only progressive direction would be to socialize the relations of production, that is, create a system of property relations through which the means of production become the common property of those who work them and rights and claims are provided for the consumers of the products. It may be necessary first to nationalize the mines and processing facilities by turning them into state property, but this would have little bearing on the relations of production (although presumably it could ameliorate the worst excesses of exporting jobs abroad). If there are to be equitable and just relations of production and a guarantee of the safest working conditions, it will be necessary for those most directly affected to control the conditions of their work. Miners will have to control the mines.

APPENDIX I

Gathering the Data

In preparing this book the first task was to examine the trade journals for the mining industry, focusing primarily on material related to technology and the labour force. This was done by a research assistant and myself. I then began to make arrangements with Inco Limited and the United Steelworkers of America (USWA) for access to information and the work sites. For the company's assistance I dealt with the Toronto head office. Here I met on several occasions the heads of personnel and public relations, and my visits to the sites had to be approved by the president and chairman. For help from the unions I contacted each level, national head office, regional offices, and the locals. The presidents of the locals were able to be of most assistance in arranging interviews (although archival material collected at the other levels was very useful). They included Dave Patterson, president of Local 6500 in Sudbury, Ray Moreau, president of Local 6200 in Port Colborne, and Dick Martin, president of Local 6166 in Thompson. After completing these contacts, I made arrangements to go to each location and see various designated people.

Before going into the field, we undertook additional archival research. I examined the briefs, exhibits, and transcripts of the Royal Commission on Health and Safety of Workers in Mines, known as the Ham Commission (deposited in the Archives of Ontario), and my research assistant, John Baker, and I went through the publications of the company and the union.

These publications proved most valuable in helping to identify specific issues in the mining industry, as an introduction to the technology of the industry and the various operations of Inco, and in providing some preliminary information on work conditions. All this was necessary orientation in the language and issues of the industry so that the optimum use could be made of the field research.

At Inco offices and plants we interviewed the employee relations or industrial relations officials, the heads of engineering, and the managers of plants and mines. From the employee relations people we learned about training, grievances, penalties, and turnovers. The engineers provided information on technology in use or being developed. They were concerned at the work sites primarily with implementing and maintaining the technology and in the laboratory primarily with development of further technology. The plant and mine managers provided more specific information on all aspects of whatever operation they were overseeing and what change had occurred; they were asked to explain changes that had taken place, how the work now differed, and particularly how the changes affected the labour required and the skill levels of those workers. The local presidents of unions and the chairmen of various committees such as grievance, health and safety, compensation, and bonus were interviewed. They provided detailed information both orally and from their records on their relations with the company and views on developments in the industry.

In the field work, my research assistant and I spent a total of thirty-three working days in Sudbury, sixteen in Thompson, and six in Port Colborne in the summer of 1977. I made an additional trip to Sudbury during the strike in March, 1979. Day trips were made to the research laboratories in Sheridan Park, the Toronto head office, and the district and national offices of USWA. The work was organized well in advance, and tours of work sites were scheduled. At all sites detailed notes were taken and information was gathered from company and union officials. Arrangements were also made for after-hours talks with workers from various operations that we met during these tours, and some interviews with recently retired employees were scheduled. The result was taped interviews with over fifty people and notes from discussions with hundreds more. Tapes were used for in-depth interviews and when conditions warranted. Obviously taping was not practical underground or in noisy areas, and there each of us kept notes.

There were tours to seven mines selected on the basis of their size, distance from town, degree of mechanization, accident rates, and degree of militancy among the workers.* These indicators were worked out with both company and union officials. Half of all Inco's mines were included. Virtually all surface operations were toured, twelve in all. Each visit took about half a day, and evenings and other days were spent interviewing. By working eighteen-hour days, the two of us were able to gather a good deal of information. Besides conducting the interviews, we received data on the number of workers per operation, ratios of production workers to maintenance workers, numbers of hourly workers, and output of each operation from company and union officials. We acquired, as well, detailed job descriptions for about two hundred occupations in the whole mining process, including the worker's training, his equipment, his supervisor(s), and his own supervisory capacity, if any.

At each site we focused on the general level of technology in use at that operation, specifically noted innovations that had been introduced, and asked workers to explain changes and conditions. This proved useful for observing the direct effects of technology. We were always accompanied by either the plant manager (usually a person with technical training) or an engineer familiar with the operation and processes. They were instructed by head office to allow us to speak without interference with anyone we wished (and we could take better advantage of the opportunity because there were two of us) and to examine any part of the operation we chose. It was readily apparent that it was a no-holds-barred arrangement, and we found good co-operation and no resistance to our questions from workers in after-hours meetings.

* Of the seven mines, one (Pipe Mine in Thompson) was an open-pit mine. The types of mines we visited, therefore, corresponded to the national average, for only about 11 per cent of the ore removed from nickel mines is from open pits, the rest coming from underground operations.

In the range of Inco mines and plants visited we were able to study both modern and traditional types of operations. In the mines, the methods used were traditional mining essentially as it has been practised since the First World War, mechanized ramp mining (where diesel equipment drives down from surface and is maintained in central underground bays), open-pit mines, and mines that are traditional except that "captured" diesel scooptrams are used in the stopes. On the surface there were highly automated mills and others requiring much more labour; there were smelters still much as they had been since 1930 and others using modern processes. The refinery practices varied from ultramodern high-pressure carbonyl processes to the traditional electrolytic process. In addition, the same types of operations and processes differed greatly from one location to another, and these differences made it necessary to see nearly all the surface operations. Needless to say, this took time, but it certainly helped in establishing the contrasts between the production methods.

Several times we considered administering a questionnaire to Inco employees as a useful supplement to the archival material, on-site tours, and in-depth interviews. Given the circumstances of layoffs and strike, however, it was not possible to do this. Inco had tentatively agreed to the procedure but withdrew permission during the disruptions that occurred from mid 1977 into 1979. A questionnaire for union members only was considered, but because of the limitations of such information and the particularly bitter relations brought on by the layoffs and strike, the results would have had dubious validity, even if the workers had been willing to respond. The main advantage of a general questionnaire would have been to gain more systematic data on workers' attitudes and perceptions. Systematic information on other factors can be obtained from non-questionnaire sources, but the questionnaire data would have been a valuable source for confirming or denying many other ideas and could have been helpful in developing a common body of information for comparison when studying other work settings.

APPENDIX II

Financial Performance of Falconbridge Nickel and Inco, 1971 to 1978

Net profit after taxes ($000)		*Year*	*Return on investment (net earnings/shareholders' equity)*	
Falconbridge	Inco		Falconbridge	Inco
17,513	94,242	1971	6.0	8.5
5,529	109,906	1972	2.1	10.2
47,904	226,859	1973	16.9	18.2
21,976	306,002	1974	9.8	21.1
3,221	186,889	1975	1.0	12.6
14,703	196,758	1976	4.6	12.6
(29,223)	99,859	1977	−7.6	5.9
5,500	77,809	1978	1.4	3.7

SOURCE: Calculated from consolidated financial statements in annual reports, Financial Post Corporation Service.

APPENDIX III

Canadian Exports of Nickel by Country and Refining Stage
(percentages)

	Nickel ore and matte			*Nickel oxides*			*Refined nickel*		
	1965	1970	1974	1965	1970	1974	1965	1970	1974
United States	—	—	—	66	62	67	82	68	64
United Kingdom	57	41	40	18	23	9	11	17	12
Norway	40	49	50	—	—	—	—	—	—
Japan	3	10	10	—	—	—	1	2	2
EEC	—	—	—	12	12	18	3	8	2
Sweden	—	—	—	1	—	2	—	—	—
Australia	—	—	—	2	2	2	1	1	1
China	—	—	—	—	—	—	—	1	17
India	—	—	—	—	—	—	1	1	1
Total	100	100	100	99	99	98	99	98	99
Tons*	74.7	88.8	85.2	37.2	39.8	51.1	112.6	139.0	118.8

SOURCE: Calculated from Energy, Mines and Resources Canada, *Nickel* (Ottawa, 1976), pp. 9-10.

* Metric tons nickel content, '000.

APPENDIX IV

Production and Non-production Workers in Canadian Mining, 1973

	Production and related workers		*Non-production*		*All employees*	
	No.	% of all employees	No.	% of all employees	No.	%
Metal mines						
Nickel, copper	14,696	73.4	5,335	26.6	20,031	21.2
Copper, gold, silver	12,994	74.0	4,577	26.0	17,571	18.6
Iron	8,521	63.6	4,874	36.4	13,395	14.2
Gold quartz	4,727	84.4	876	15.6	5,603	5.9
Silver, lead, zinc	4,489	73.4	1,623	26.6	6,112	6.5
Misc. metals	2,557	74.7	865	25.3	3,422	3.6
Total	47,984	72.6	18,150	27.4	66,134	69.9
Non-metal mines						
Asbestos	6,430	80.1	1,597	19.9	8,027	8.5
Coal	6,445	82.0	1,411	18.0	7,856	8.3
Stone	2,590	83.6	507	16.4	3,097	3.3
Sand and gravel	1,631	74.9	548	25.1	2,179	2.3
Misc.	5,681	77.1	1,683	22.9	7,364	7.8
Total	22,777	79.9	5,746	20.1	28,523	30.1
TOTAL	70,761	74.8	23,896	25.2	94,657	100.0

SOURCE: Calculated from Statistics Canada, *Mineral Industries: Principal Statistics, 1973*, Cat. No. 26-204, Annual (Ottawa: Information Canada, 1976).

APPENDIX V

Occupations in Canadian Metal Mines: 1971 Census

	Total*	%	Male	Female
Managerial, Administrative	1,630	2.4	1,575	55
Natural Science, Engineering, and Mathematics	5,570	8.2	5,420	150
Physical science	(2,085)		(1,975)	(105)
Architects and engineers	(1,755)		(1,745)	(5)
Other occupations in engineering	(1,510)		(1,480)	(25)
Clerical and related	4,340	6.4	2,290	1,025
Stenography and typing	(1,160)		(45)	(1,115)
Bookkeeping, accounting	(1,055)		(745)	(315)
Bookkeeping and accounting clerks	(955)		(670)	(285)
Material recording, scheduling and distributing	(825)		(805)	(20)

APPENDIX V Continued

	Total*	%	Male	Female
Service occupations	1,395	2.1	1,250	150
Mining and quarrying	28,080	41.3	27,945	135
Foremen	(4,330)		(4,330)	(5)
Rock and soil drillers	(5,350)		(5,350)	(. . .)
Cutting, handling, loading	(9,630)		(9,630)	(. . .)
Processing occupations	3,735	5.5	3,700	35
Mineral ore treating	(2,995)		(2,965)	(30)
Machining	2,495	3.7	2,485	10
Metal machinist	(770)		(760)	(5)
Machinist	(505)		(495)	(5)
Metal shaping and forming	(1,600)		(1,600)	(5)
Welding	(1,280)		(1,280)	(5)
Product fabricating, Assembling, and Repair	6,655	9.8	6,635	25
Mechanics and repairers	(6,090)		(6,070)	(20)
Industrial mechanics	(4,485)		(4,460)	(20)
Construction trades	4,385	6.4	4,355	25
Excavating, grading, paving	(600)		(600)	(. . .)
Electric power, lighting, wire communication erecting	(1,980)		(1,970)	(10)
Electricians	(1,645)		(1,635)	(10)
Other construction trades	(1,800)		(1,785)	(15)
Carpenters	(670)		(665)	(5)
Plumbers	(570)		(560)	(5)
Transport equipment operating	3,135	4.6	3,110	20
Motor transportation operator	(1,735)		(1,370)	(5)
Truck drivers	(1,690)		(1,680)	(5)
Other	(1,160)		(1,145)	(15)
Materials handling	2,735	4.0	2,725	10
Hoisting	(1,315)		(1,315)	(. . .)
Materials equipment operator	(860)		(860)	(. . .)
Other crafts and equipment operators	835	1.2	825	10
Other	800	1.2	715	85
Not classified	970	1.4	965	5
Not stated	1,300	1.9	1,240	60
Total	68,050	100	65,225	2,825

SOURCE: Calculated from Statistics Canada, Special Bulletin 1971 Census of Canada, *Economic Characteristics: Occupation by Industry*, Cat. No. 94-792 (SE-1) (Ottawa: Information Canada, May 1976), Table 2.

* All occupations with five hundred or more individuals are included. Occupations in parentheses are sub-sets of the main categories with over five hundred. Male and female do not always exactly equal the total because of census rounding to the nearest five.

APPENDIX VI

Annual Labour Costs per Employee in Canadian Metal Mining (1972) and Durables Manufacturing (1971)

	Mining		*Manufacturing*	
	$	As per cent of basic pay	$	As per cent of basic pay
Pay for time worked:				
Basic pay	7,001	100.0	6,279	100.0
Overtime	350	5.0	311	5.0
Premium:				
Underground production bonus	688	9.8	...	...
Other	275	3.9	256	4.1
Paid absence:	831	11.9	673	10.1
Holiday pay	260	3.7	257	4.1
Vacation pay	565	8.1	397	6.3
Other	5	0.1	19	0.3
Misc. direct payment	152	2.2	76	1.2
Total (gross payroll)	9,298	132.8	7,595	121.0
Employer contributions:				
Pension plans	282	4.0	296	4.7
Health and medicare	187	2.7	233	3.7
Other	353	5.0	212	3.4
Total Employer Contributions	822	11.7	741	11.8
TOTAL (wage package)	10,120	144.6	8,336	132.8

SOURCE: Energy, Mines and Resources Canada, *Mining and Manpower* (Ottawa, 1976), p. 28.

APPENDIX VII

Hourly Wage Rates for
Selected Canadian Mining Occupations, October 1973
($ per hour)

Underground	Number employed	Average wage
Cage man	410	4.38
Car dropper	686	4.46
Dinkey-engine operator	191	4.17
Grizzly worker	319	4.54
Hoist operator	402	4.76
Labourer, underground	1,847	4.19
Mechanical-shovel operator	342	4.37
Miner, all-round	6,280	4.48
Miner, helper	705	3.73
Timber and steel-prop	529	4.45
Track repairman	143	4.25
Surface and mill		
Bit-sharpener tender	47	4.11
Blacksmith	55	4.67
Carpenter, maintenance	284	4.62
Crusher tender	421	4.20
Diesel mechanic	673	5.02
Electrical repairman	811	4.89
Filtering attendant	183	4.17
Flotation-cell tender	238	4.34
Grinding and classifier	322	4.35
Labourer, non-production	710	3.69
Leaching operator	53	4.47
Maintenance machinist	314	4.83
Maintenance-man helper	442	4.02
Millwright	761	4.84
Pipe fitter, maintenance	150	4.64
Truck driver	583	4.38
Welder, maintenance	461	4.86

SOURCE: Energy, Mines and Resources Canada, *Mining and Manpower* (Ottawa, 1976), p. 49.

APPENDIX VIII

Inco's Ontario Operations

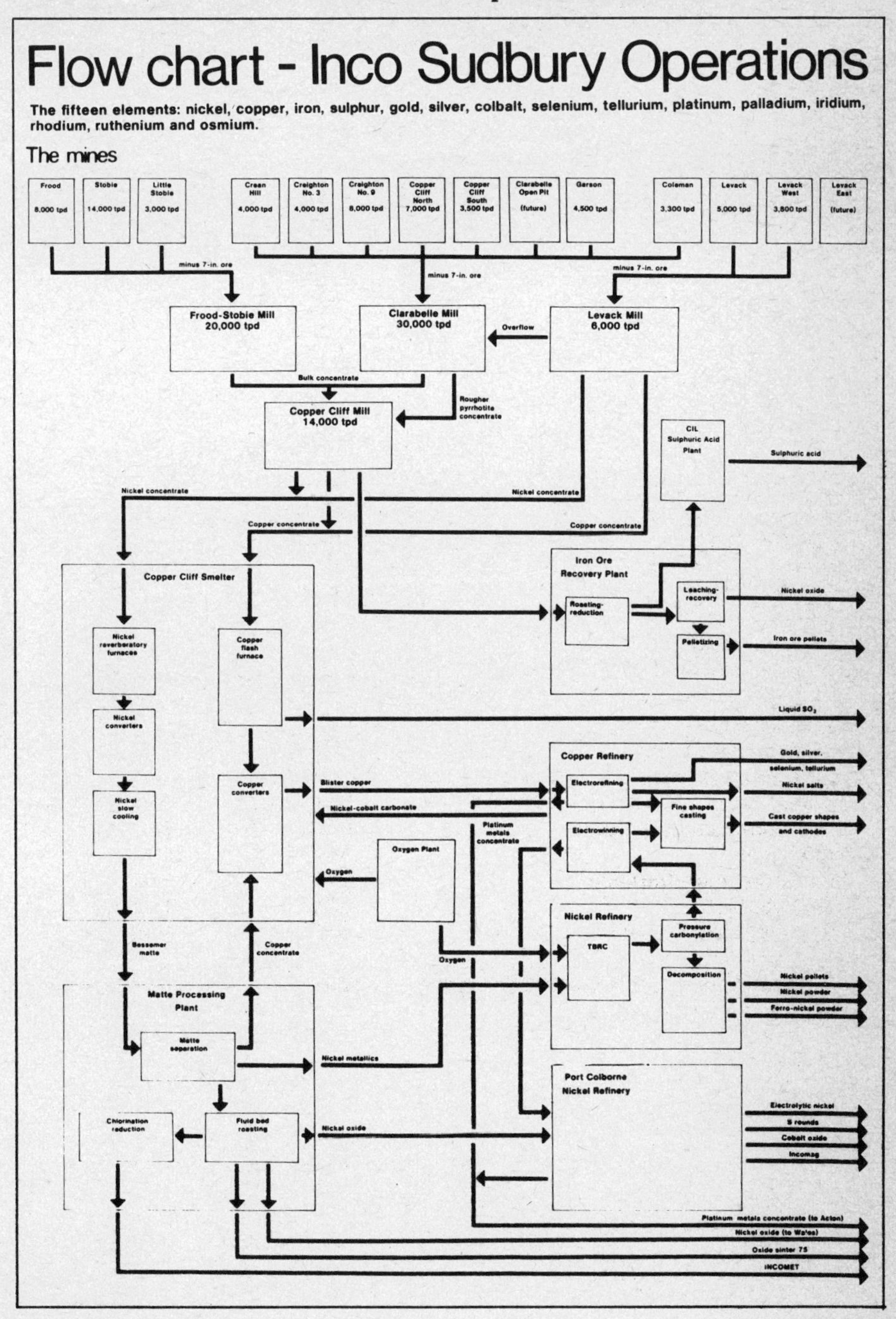

Canadian Mining Journal (May 1977):62.

APPENDIX IX

Inco's Product Flow

Smelters	Product	Destination
Copper Cliff	nickel oxide	Market: stainless steel, construction alloys, iron and steel castings
	nickel sulphide matte concentrate	Japanese refineries, Port Colborne, and Copper Cliff refineries for processing
	nickel oxide matte concentrate	Inco Clydach, U.K., and Port Colborne nickel refineries
	magnetic matte concentrate	Copper Cliff Nickel Refinery
	blister copper	Copper Cliff Copper Refinery
	sulphur dioxide	Adjacent Canadian Industries plant for processing
Sudbury Iron Ore Recovery Plant	iron ore pellets	Market
	nickel oxide powder	Nickel plating market
	sulphur dioxide	Adjacent Canadian Industries plant for processing
Thompson	cupriferrous nickel matte	Thompson refinery and Japanese refineries for processing

Refineries	Product	Destination
Port Colborne	electrolytic nickel	Market: nickel plating, stainless steel, high-nickel alloys, iron and steel castings, construction alloys, copper and brass products
	foundry additives	Iron and steel castings markets
	metal anode residues	Copper Cliff Nickel Refinery
	copper residue	Copper Cliff Smelter
	cobalt oxide	Inco Clydach cobalt refinery
Thompson	electrolytic nickel	Market: nickel plating, stainless steel, high-nickel alloys, iron and steel castings, construction alloys, copper and brass products
	sulphur	Market
	cobalt oxide	Inco Clydach cobalt refinery
	copper residues	Copper Cliff Smelter
Copper Cliff (precious metal refinery)	gold, silver	Semi-refined to market
	platinum metals	To platinum metal refinery, Acton, England
Copper Cliff (selenium and tellurium)	selenium, tellurium, and nickel sulphate	Market

SOURCE: Energy, Mines and Resources Canada, *Nickel* (Ottawa, 1976), pp. 14-16.

APPENDIX X

Fatalities in Canadian Industries by Industrial Divisions, 1967 to 1976 *(annual averages)*

	Agri-culture	Forestry	Fishing	Mining	Manu-facturing	Con-struction	Trans-port	Trade	Finance	Service	Public Admin.	All
Fatalities*	25	87	19	167	215	215	218	73	4	72	67	1,162
Workers ('000)	504	76	23	125	1,866	529	735	1,420	394	2,159	545	8,376
Incidence rates†	5.0	114.5	82.6	133.5	11.5	40.6	29.7	5.1	1.1	3.3	12.3	13.9

SOURCE: *Labour Gazette*, Dec. 1977, Tables 1, 3, 4, pp. 557-58.

*Fatalities "occurred to workers during the course of, or arose out of, their employment. They include deaths of pensioners suffering from industrial diseases such as silicosis and lung cancer as well as deaths of pensioners who suffered earlier disabling work injuries."

† Fatality incidence rate: number of cases per 100,000 workers.

APPENDIX XI

Fatal Injuries and Compensable Injuries in Ontario Mining, 1970 to 1975

Fatal Injuries:

	Man-hours worked (millions)	Man-years at risk	Fatalities	Fatalities per 10,000 man-years at risk	Fatalities per million man-hours worked
1970	86.5	43,250	24	5.5	0.28
1971	82.6	41,300	22	5.3	0.27
1972	71.4	35,700	14	3.9	0.20
1973	66.8	33,400	11	3.3	0.16
1974	69.6	34,800	16	4.5	0.23
1975	69.5	34,750	9	2.6	0.13

Compensable Injuries:

	Injuries	Injuries per 100 man-years at risk	Injuries per million man-hours worked
1970	3,575	8.2	41.3
1971	3,318	8.0	40.1
1972	2,909	8.1	40.7
1973	3,220	9.6	48.2
1974	3,747	10.8	53.7
1975	4,246	12.2	61.0

SOURCE: Royal Commission on the Health and Safety of Workers in Mines, *Report* [the Ham Commission] (Toronto: Ministry of the Attorney General of Ontario, 1976), pp. 122, 126.

APPENDIX XII

Number of Workers per Lost Time Accident in Thompson, 1965 to 1976*

	Thompson Mine	Birchtree Mine	Pipe Mill	Mill	Smelter	Refinery	Maintenance
1965	22	...	...	39	26	22	13
1966	24	...	...	0	33	21	19
1967	28	...	...	0	32	16	30
1968	19	32	...	31	104	25	28
1969	25	11	...	34	36	46	30
1970	42	22	...	82	113	142	28
1971	25	24	30	28	34	40	18
1972†	5	6	31	13	6	8	5
1973	4	5	16	10	6	7	8
1974	5	4	11	6	6	6	10
1975	6	5	6	10	5	4	10
1976	5	4	4	9	9	7	12
1965-76 average	17.5	12.6	16.3	24.7‡	34.2	28.7	17.6

* The figures are derived by dividing the number of employees by the number of lost time accidents. If every employee were injured in a year the figure would be one, if every fifth employee, the number would be five.

† After 1972 Inco policy was that "employees who were not able to perform their regular duties should not report to work, as it has become increasingly difficult to find suitable job openings for employees who were incapacitated in some way" (private communication).

‡ Post 1967.

APPENDIX XIII

Distribution of Hourly Labour Force in Thompson

	Thompson Mine	Birchtree Mine	Pipe Mine	Soab Mine	*Total Mines*	Mill	Smelter	Refinery	*Total Surface*	Maintenance	*Ratio* Production: Maintenance	*Total* %	*Total* (N)
1965	43.7	...	...	...	43.7	2.3	13.6	21.4	37.3	19.0	4.3 : 1	100	1,712
1966	46.4	...	...	...	46.4	2.4	14.0	21.3	37.7	16.0	5.3 : 1	100	1,663
1967	48.0	...	...	...	48.0	2.3	14.2	19.6	36.1	16.0	5.3 : 1	100	2,062
1968	48.5	1.4	...	...	49.9	2.6	13.1	17.7	33.4	16.8	5.0 : 1	100	2,374
1969	42.7	11.4	...	...	54.1	2.4	11.2	14.5	28.1	17.8	4.6 : 1	100	2,877
1970	40.1	14.7	0.1	...	54.9	2.5	10.4	13.0	25.9	19.1	4.2 : 1	100	3,259
1971	34.8	13.9	2.6	1.8	53.1	3.3	10.8	12.9	27.0	20.0	4.0 : 1	100	3,436
1972	33.9	12.9	1.0	...	47.9	2.3	12.6	16.0	30.9	21.2	3.7 : 1	100	2,843
1973	33.1	11.7	3.4	...	48.2	2.2	12.5	15.1	29.8	22.0	3.5 : 1	100	2,851
1974	33.5	10.8	5.6	...	49.9	2.2	12.8	13.1	28.1	22.0	3.6 : 1	100	2,857
1975	34.6	12.5	6.4	...	53.5	2.2	12.6	9.4	24.2	22.4	3.5 : 1	100	2,796
1976	36.0	13.3	3.6	...	52.9	2.1	13.1	9.3	24.5	22.7	3.4 : 1	100	2,581
1977	36.0	11.7	4.2	...	51.9	2.1	13.7	8.9	24.7	23.5	3.3 : 1	100	2,600

NOTES

CHAPTER ONE

1. H. A. Innis, *The Fur Trade in Canada: An Introduction to Canadian Economic History* (1930), rev. ed. (Toronto: University of Toronto Press, 1956), p. 385.
2. H. A. Innis, "The Canadian Economy and the Depression" (1934), in *Essays in Canadian Economic History*, ed. Mary Q. Innis (Toronto: University of Toronto Press, 1956), p. 138.
3. H. A. Innis, "Government Ownership and the Canadian Scene" (1933) in M. Q. Innis, ed., *Essays*, p. 88.
4. H. A. Innis, *Settlement and the Mining Frontier*, in Canadian Frontiers of Settlement, ed. W. A. Mackintosh and W. L. G. Joerg (Toronto: Macmillan, 1936), p. 171.
5. Ibid., p. 402.
6. H. A. Innis and B. Ratz, "Labour," in *Encyclopedia of Canada*, ed. W. S. Wallace (Toronto: University Association of Canada, 1940), vol. 3, p. 363.
7. See Wallace Clement, *Continental Corporate Power: Economic Elite Linkages between Canada and the United States* (Toronto: McClelland and Stewart, 1977).
8. See Erik Olin Wright, *Class, Crisis and the State* (London: New Left Review Editions, 1978), p. 37.
9. Karl Marx, *Capital*, vol. 3 (1894; reprint ed., New York: International Publishers, 1967), p. 884.
10. See Guglielmo Carchedi, "Reproduction of Social Classes at the Level of Production Relations," *Economy and Society* 4 (Feb. 1975):14-16; see also his *On the Economic Identification of Social Classes* (London: Routledge and Kegan Paul, 1977).
11. Marx, *Capital*, vol. 1, (1867; reprint ed., New York: International Publishers, 1967), p. 364.
12. Ibid., p. 361.
13. Harry Braverman, *Labor and Monopoly Capital: The Degradation of Work in the Twentieth Century* (New York: Monthly Review Press, 1974), p. 188.

CHAPTER TWO

1. See "Mining Techniques in the Yukon," reprinted from William B. Haskell, *Two Years in the Klondike and Alaska Gold-fields* (Hartford: 1898), in *Let Us Be Honest and Modest: Technology and Society in Canadian History*, eds. B. Sinclair, N. R. Ball, and J. O. Peterson (Toronto: Oxford University Press, 1974), pp. 197-98.
2. See H. A. Innis, *Settlement and the Mining Frontier*, in Canadian Frontiers of Settlement, ed. W. A Mackintosh and W. L. G. Joerg (Toronto: Macmillan, 1936), pp. 199-203.
3. William Moore, "Staples and the Development of the Capitalist Mode of Production" (Master's thesis, Department of Sociology, McMaster University, Hamilton, 1978), p. 203.
4. Innis, *Mining Frontier*, pp. 198, 207.
5. A. N. C. Threadgold, quoted in Innis, *Mining Frontier*, p. 207n.
6. See Innis, *Mining Frontier*, p. 207.
7. Ibid., p. 259.
8. Ibid., p. 176.
9. Ibid., p. 197.
10. See ibid., pp. 226-27.
11. Quoted in ibid., p. 228.
12. Ibid., p. 268.
13. Ibid., pp. 223-24.
14. See ibid., pp. 238-39.
15. See Paul M. Koroscil, "Robert Dunsmuir: A Portrait of a Western Capitalist," in *Canadian Frontier*, ed. P. M. Koroscil (Surrey, B.C.: Antonson Publishing, 1978), p. 15.
16. See Gustavus Myers, *A History of Canadian Wealth* (1914; reprint ed., Toronto: James Lorimer, 1975), pp. 302-3.
17. See ibid., pp. 306-7.
18. Moore, "Capitalist Mode," pp. 200-01.
19. See Greg Kealey, ed., *Canada Investigates Industrialism* (Toronto: University of Toronto Press, 1973), pp. 404-42.
20. See S. D. Hanson, "Estevan 1931," in *On Strike: Six Key Labour Struggles in Canada, 1919-1949*, ed. Irving Abella (Toronto: James Lorimer, 1974), pp. 33-35, 66 and 70; more generally, see Stuart Jamieson, *Times of Trouble: Labour Unrest and Industrial Conflict in Canada, 1900-66*, Study No. 22, Task Force on Labour Relations (Ottawa: Privy Council Office, October 1968), p. 96.
21. See Jamieson, *Trouble*, pp. 95, 197; there is also an account of some of these strikes in H. A. Innis and B. Ratz, "Labour," in *Encyclopedia of Canada*, ed. W. S. Wallace (Toronto: University Association of Canada, 1940), vol. 3, p. 359.
22. See Moore, "Capitalist Mode," p. 205.
23. Innis, *Mining Frontier*, p. 313; also, see E. S. Moore, *American Influence in Canadian Mining* (Toronto: University of Toronto Press, 1941), pp. 72-73.
24. Quoted in Innis, *Mining Frontier*, pp. 294-95.
25. See ibid., p. 314.
26. Ibid., p. 299.
27. See *Report* of the Royal Commission on Dominion-Provincial Relations, Book I (Ottawa: King's Printer, 1940), p. 146.
28. Innis, *Mining Frontier*, p. 316.

29. Stanley Scott, "A Profusion of Issues: Immigrant Labour, the World War, and the Cominco Strike of 1917," *Labour/Le Travailleur* 2 (1977):58.
30. See ibid., p. 59.
31. Ibid., p. 64.
32. Reprinted in Sinclair, Ball, and Peterson, eds., *Honest and Modest*, p. 205.
33. Doug Baldwin, "A Study in Social Control: The Life of the Silver Miner in Northern Ontario," *Labour/Le Travailleur* 2 (1977):82.
34. Innis, *Mining Frontier*, p. 326.
35. Baldwin, "Social Control," pp. 85-86.
36. Ibid., p. 87.
37. H. A. Innis, "Recent Developments in the Canadian Economy" (1941) in *Essays in Canadian Economic History*, ed. Mary Q. Innis (Toronto: University of Toronto Press, 1956), p. 299.
38. See Innis, *Mining Frontier*, pp. 360-62.

CHAPTER THREE

1. See E. S. Moore, *American Influence in Canadian Mining* (Toronto: University of Toronto Press, 1941), p. 28.
2. O. W. Main, "International Nickel: The First Fifty Years," in *Canadian Business History: Selected Studies, 1497-1971*, ed. David S. Macmillan (Toronto: McClelland and Stewart, 1972), p. 255; see also O. W. Main, *The Canadian Nickel Industry: A Study in Market Control and Public Policy* (Toronto: University of Toronto Press, 1955), p. 95.
3. See Wallace Clement, *Continental Corporate Power: Economic Elite Linkages between Canada and the United States* (Toronto: McClelland and Stewart, 1977), pp. 154-55.
4. H. V. Nelles, *The Politics of Development: Forests, Mines and Hydro-electric Power in Ontario, 1849–1941* (Toronto: Macmillan, 1974), p. 328.
5. See Main, *Canadian Nickel*, p. 84.
6. See John Deverell, *Falconbridge: Portrait of a Canadian Mining Multinational* (Toronto: James Lorimer, 1975), pp. 39-50.
7. See *Canadian Mining and Metallurgical Bulletin* 67 (1964):3.
8. John F. Thompson and Norman Beasley, *For the Years to Come: A Story of International Nickel of Canada* (New York: G. P. Putnam's Sons, 1960), p. 354.
9. Gilbert A. Stelter, "Community Development in Toronto's Commercial Empire: The Industrial Towns of the Nickel Belt," *Laurentian University Review* 6, no. 3 (1974):6.
10. See ibid., p. 7.
11. See Deverell, *Falconbridge*, p. 90.
12. See Gilbert Stelter, "The Origins of a Company Town: Sudbury in the Nineteenth Century," *Laurentian University Review* 3, no. 3 (1971):4-5.
13. See *Directory of Industries – Regional Municipality of Sudbury* (Sudbury Regional Development Corporation, Nov. 1974), p. 25.
14. *Report* of Joint Manpower Adjustment Committee at Canadian Furnace Division of the Algoma Steel Corporation Ltd. at Port Colborne, Ontario (June 1977), p. 9.
15. See *Thompson Citizen*, Feb. 1978:1.
16. Jamie Swift, *The Big Nickel: Inco at Home and Abroad* (Kitchener: Between the Lines Press, 1977), p. 94.
17. House of Commons *Debates* 121:5 (24 Oct. 1977):159.

18. See *Financial Post*, 5 Nov. 1977:11.
19. Quoted in *Toronto Star*, 29 Oct. 1977:A3.
20. Inco Limited, *Annual Report for 1976* (Feb. 1977), p. 11.

CHAPTER FOUR

1. See P. E. Nickel et al., *Economic Impacts and Linkages of the Canadian Mining Industry* (Kingston: Queen's University, Centre for Resource Studies, 1978), p. 22, Table 5.
2. See ibid., p. 23.
3. See Bruce W. Wilkinson, "Trends in Canada's Mineral Trade," in *Working Paper No. 9* (Kingston: Queen's University, Centre for Resource Studies, 1978), pp. 6, 17.
4. Nickel et al., *Economic Impacts*, p. 24n.
5. See Wilkinson, "Trends," p. 16.
6. See Nickel et al., *Economic Impacts*, pp. 88-89.
7. See ibid., p. 102.
8. John Deverell, *Falconbridge: Portrait of a Canadian Mining Multinational* (Toronto: James Lorimer, 1975), p. 11.
9. See *Financial Post*, 2 Oct. 1976:45.
10. See *Canadian Mining Journal* 96, no. 2 (1975):67; 97, no. 2 (1976):75.
11. See Inco Limited, *Annual Report for 1977* (Feb. 1978), p. 2.
12. See *Financial Post*, 10 Feb. 1979:26.
13. See *Globe and Mail*, 13 Feb. 1979:B1.
14. See *Globe and Mail*, 27 Feb. 1979:B10.
15. *Globe and Mail*, 15 Oct. 1977:B13.
16. See Royal Commission on Canada's Economic Prospects, *Final Report* (Ottawa: Queen's Printer, 1955), p. 92.
17. See *Mine-Mill Herald*, June 1961:4.
18. *Northern Miner*, 6 June 1974:54.
19. *Canadian Mining Journal* 67, no. 5 (1946):501.
20. See Energy, Mines and Resources Canada, *Mining Communities* (Ottawa: 1976), p. 8.
21. See Energy, Mines and Resources Canada, *Mineral Industry Trends and Economic Opportunities* (Ottawa: 1976), p. 27.
22. J. A. MacMillan, G. S. Gislason, and S. Lyon, *Human Resources in Canadian Mining* (Kingston: Queen's University, Centre for Resource Studies, 1977), p. xiv.
23. See ibid., pp. 7, 14.
24. Ontario Royal Commission on Nickel, *Report* (Toronto: King's Printer, 1917), p. 225.

CHAPTER FIVE

1. See J. A. MacMillan, G. S. Gislason, and S. Lyon, *Human Resources in Canadian Mining* (Kingston: Queen's University, Centre for Resource Studies, 1977), pp. 28-29.
2. *Engineering and Mining Journal*, Sept. 1975:91-92.
3. *Northern Miner*, 7 Aug. 1975:13.
4. *Engineering and Mining Journal*, June 1976:150.
5. *Canadian Mining Journal* 98, no. 5 (1977):17.
6. Ibid.
7. See *Engineering and Mining Journal*, Sept. 1975:108.
8. See ibid., p. 104.

9. Ibid., p. 101.
10. *Engineering and Mining Journal*, June 1976:162.
11. Earle A. Ripley, Robert E. Redmann, and James Maxwell, *Environmental Impact of Mining in Canada* (Kingston: Queen's University, Centre for Resource Studies, 1978), p. 35.
12. *Canadian Mining Journal* 95, no. 5 (1974):18.
13. *Canadian Mining Journal* 98, no. 5 (1977):14.
14. *Inco Triangle* 33, no. 6 (1973):4.
15. E. Trist and K. Bamforth, "Some Social and Psychological Consequences of the Longwall Method of Coal Getting," *Human Relations* 4 (1951):6.
16. Ibid., p. 8.
17. Ibid., pp. 9-10.

CHAPTER SEVEN

1. See *Miners' Voice*, Oct. 1977:12.
2. Royal Commission on Health and Safety of Workers in Mines [the Ham Commission] *Report* (Toronto: Ministry of the Attorney General [Ontario], 1976), pp. 145-46.
3. Ray Moreau, Local 6200 USWA, Ham Commission Brief 109 (18 Feb. 1975):1-2.
4. *Miners' Voice*, May 1978:13.
5. See Ham Commission *Report*, pp. 132-38, 144, Appendix Table D.8.
6. K. Moisey, Co-chairman, Inquest Committee Local 6500 USWA, Ham Commission Brief 59 (30 Jan. 1975):1.
7. J. Hickey, Compensation and Welfare Officer, Local 6500 USWA, Ham Commission Brief 62 (30 Jan. 1975):4-5.
8. See Ham Commission *Report*, pp. 225-26.
9. Ibid., p. 10.
10. *Globe and Mail*, 6 Oct. 1976:57.
11. C. Lambert, Local 6500 USWA, Ham Commission Brief 50 (27 Jan. 1975):22.
12. C. Hews, Inco, before the Ham Commission, Transcript (28 Jan. 1975):2099.
13. See *Globe and Mail*, 10 Sept. 1976:8.
14. Ibid.
15. *Steel Labor* 42, no. 6 (1977):4.
16. See *Miners' Voice*, Dec. 1976:6.
17. *Miners' Voice*, Feb. 1979:7.
18. See *Globe and Mail*, 18 May 1977:5.
19. *Miner's Voice*, July 1978:14.
20. C. Lambert, Local 6500 USWA, before the Ham Commission, Transcript (27 Jan. 1975):1968-69.
21. Ibid., pp. 23-24.
22. *Globe and Mail*, 24 April 1978:3.
23. C. Lambert, Local 6500 USWA, Ham Commission Brief 50 (27 Jan. 1975):27.
24. Ham Commission *Report*, p. 5.
25. John Deverell, *Falconbridge: Portrait of a Canadian Mining Multinational* (Toronto: James Lorimer, 1975), p. 106.
26. Ham Commission *Report*, p. 37.
27. Ibid., p. 38.

28. C. Lambert, Local 6500 USWA, before the Ham Commission, Transcript (28 Jan. 1975):1995.
29. Ham Commission *Report*, pp. 40-41.
30. Ibid., p. 46.
31. USWA, *Directors Report*, May 1977:32-33.
32. See *Globe and Mail*, 17 Jan. 1978:5.
33. See *Globe and Mail*, 6 Dec. 1977:1.
34. See *Globe and Mail*, 7 Feb. 1979:9.

CHAPTER EIGHT

1. Harry Braverman, *Labor and Monopoly Capital: The Degradation of Work in the Twentieth Century* (New York: Monthly Review Press, 1974), pp. 113-14, 119.
2. Andrew Friedman, *Industry and Labour: Class Struggle at Work and Monopoly Capitalism* (London: Macmillan, 1977), p. 82.
3. Ibid., p. 83.
4. Ibid., pp. 84-85.
5. *Inco Industrial Relations Handbook* (Sudbury: n.d.), p. 19.
6. *Financial Post*, 9 Dec. 1978:16.
7. *Globe and Mail*, 18 April 1978:8.
8. C. Hews, Inco, before the Ham Commission, Transcript (28 Jan. 1975):2077-99.
9. Colin Lambert, Local 6500 USWA, Ham Commission Brief 50 (27 Jan. 1975):8.
10. Energy, Mines and Resources Canada, *Mining and Manpower* (Ottawa, 1976), p. 15.
11. Dave Patterson, "The Education of a Nickel Miner," *Ontario Report* 2, no. 2 (1977):17.
12. *IN Manitoba* 8, no. 3 (1977):18.
13. Management Training Systems, *Inco Consolidated Report*, Ontario Division, April 1976:3.
14. Ibid., p. 6.
15. Ibid., p. 19.
16. *Collective Bargaining Agreement for Thompson between Inco Ltd. and Local 6166 USWA* (Thompson, Man.: 1976), p. 6, Article 2.02.
17. Management Training Systems, *Report*, p. 9.
18. Ibid., p. 9.
19. James R. Bright, *Automation and Management* (Cambridge: Harvard University Press, 1958), p. 183.
20. Robert Blauner, *Alienation and Freedom: The Factory Worker and His Industry* (Chicago: University of Chicago Press, 1964), p. 144.
21. Bright, *Automation*, p. 187.
22. Katherine Stone, "The Origins of Job Structure in the Steel Industry," *Review of Radical Political Economics* 6, no. 2 (1974-75):147.

CHAPTER NINE

1. Gilbert Stelter, "The Origins of a Company Town: Sudbury in the Nineteenth Century," *Laurentian University Review* 3, no. 3 (1971):27.
2. John Lang, "A Lion in a Den of Daniels: A History of the International Union of Mine, Mill and Smelter Workers in Sudbury, Ontario, 1942-1962" (Master's thesis, University of Guelph, 1970), pp. 29-31. This thesis is drawn upon heavily for the historical background of unions at Inco and supplemented by contemporary accounts.

3. Ibid., pp. 42-43.
4. See Irving Abella, *Nationalism, Communism, and Canadian Labour: The CIO, the Communist Party, and the Canadian Congress of Labour 1935-1956* (Toronto: University of Toronto Press, 1973), p. 90.
5. Lang, "Lion," p. 76.
6. Ibid., pp. 89-90.
7. Abella, *Canadian Labour*, p. 110.
8. Lang, "Lion," p. 133.
9. Ibid., p. 136.
10. Ibid., p. 139.
11. See ibid., p. 177.
12. Ibid., p. 181.
13. Ibid., p. 184.
14. See ibid., p. 201.
15. Ibid., p. 206.
16. Ibid., p. 260.
17. Ibid., p. 298.
18. *Steel Labor* 26, no. 12 (1961):4.
19. See *Mine-Mill Herald*, Oct. 1960:1.
20. Lang, "Lion," pp. 319-20.
21. See Labour Canada, *1976 Working Conditions in Canadian Industry*, Report No. 20 (Ottawa: 1977).
22. *Steel Labor* 39, no. 7 (1974):7.
23. *Collective Bargaining Agreement for Thompson between Inco Ltd. and Local 6166 USWA* (Thompson, Man.: 1976), p. 3.
24. *Collective Bargaining Agreement between Inco Ltd. and United Steelworkers of America, Local 6500*, March 1, 1976, Article 6.D1.
25. See *Steel Gauntlet*, May 1977.
26. See *Steel Labor* 29, no. 9 (1964):1.
27. *Globe and Mail*, 15 Sept. 1978:1-2.
28. *Globe and Mail*, 2 Nov. 1978:3.
29. *Globe and Mail*, 27 Sept. 1978:10.
30. *Globe and Mail*, 7 Dec. 1978:7.
31. *Globe and Mail*, 1 Dec. 1978:7.
32. *Financial Post*, 9 Dec. 1978:19.
33. *Globe and Mail*, 19 Oct. 1978:11.
34. *Globe and Mail*, 10 Oct. 1978:8.
35. See *Globe and Mail*, 5 April 1979:8.
36. *Toronto Star*, 15 March 1979:A12.
37. *Globe and Mail*, 12 April 1979:B1.
38. See *Globe and Mail*, 23 July 1979:B5.
39. *Inco Industrial Relations Handbook* (Sudbury: n.d.), pp. 16-17.
40. See Mining Association of Canada, *Labour Turnover and Shortages in the Canadian Mining Industry: An Analysis of the Principal Statistics* (Ottawa: Mining Association of Canada, 1974).
41. Energy, Mines and Resources Canada, *Mining Communities* (Ottawa, 1976), p. 8.
42. *Canadian Mining Journal* 95, no. 7 (1974):16.
43. Mike Geravelis, Local 6166 USWA, before the Ham Commission, Transcript (17 Feb. 1975):3434-37.

CHAPTER TEN

1. See Pierre L. Bourgault, *Innovation and the Structure of Canadian Industry*, Background Study No. 23 (Ottawa: Science Council of Canada, 1972).
2. See John N. H. Britton and James M. Gilmour, *The Weakest Link: A Technological Perspective on Canadian Industrial Underdevelopment*, Background Study No. 43 (Ottawa: Science Council of Canada, 1978).
3. See Britton and Gilmour, *Weakest Link*, p. 81.
4. E. S. Moore, *American Influence in Canadian Mining* (Toronto: University of Toronto Press, 1941), p. 90.
5. Britton and Gilmour, *Weakest Link*, p. 94.
6. *Canadian Mining Journal* 95, no. 7 (1974):15.
7. P. R. Richards, M. R. Leenders, C. Doucet, and P. Kuhlmann, *The Role of Innovation in the Mining and Mining Supply Industries* (Ottawa: Energy, Mines and Resources Canada, 1976), pp. 1, 56.
8. *Northern Miner*, 15 April 1976:17.
9. *Northern Miner*, 12 Feb. 1976:13.
10. Ross Hynes, "Provincial Mineral Policies: Ontario, 1945-75," in *Working Paper No. 5* (Kingston: Queen's University, Centre for Resource Studies, 1978), p. 2.
11. Ibid., pp. 16-17, 19.
12. *Globe and Mail*, 17 March 1978:5.
13. Ibid.
14. *Financial Post*, 24 Feb. 1979:M16; *Globe and Mail*, 17 May 1979:B5.
15. Hynes, "Mineral Policies," p. 12.
16. *Financial Post*, 24 Feb. 1979:M1.
17. E. Trist and K. Bamforth, "Some Social and Psychological Consequences of the Longwall Method of Coal Getting," *Human Relations* 4 (1951):17.
18. Katherine Stone, "The Origins of Job Structure in the Steel Industry," *Review of Radical Political Economics* 6, no. 2 (1974-75):123.
19. Ibid., p. 114.
20. Karl Marx, *Capital*, vol. 1 (1867; reprinted New York: International Publishers, 1967), p. 429.

Index

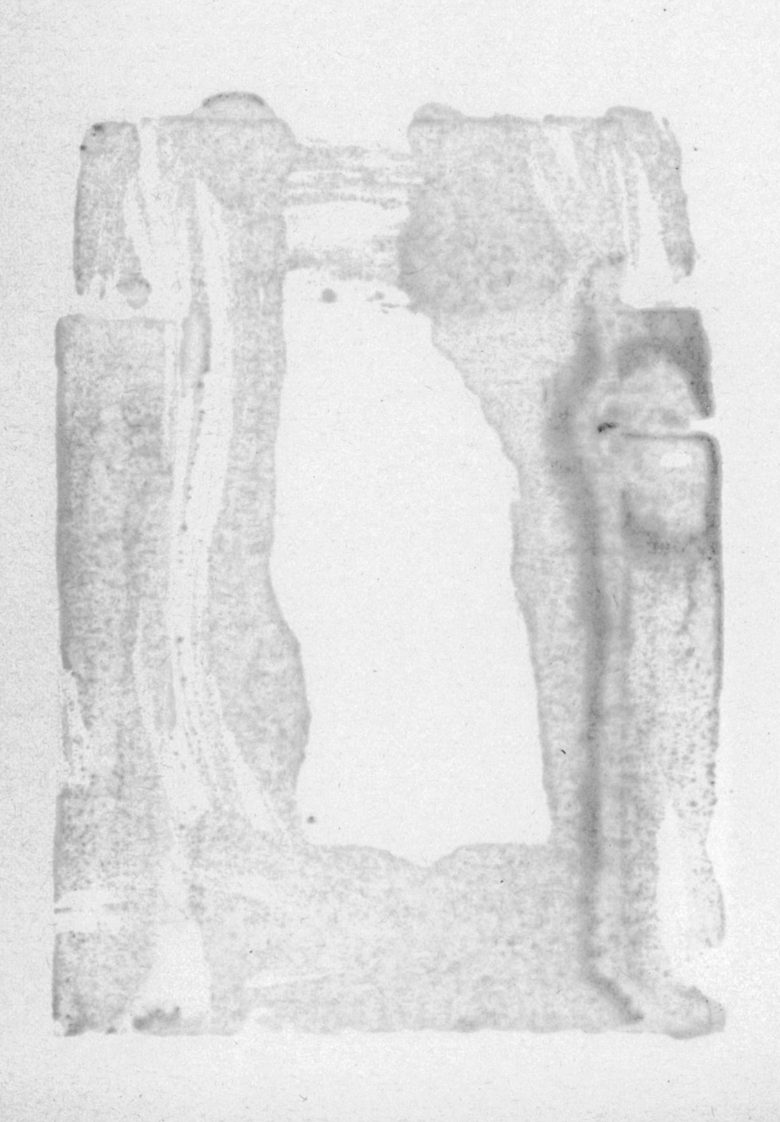